ROAD&TRACK

PETER EGAN "AT LARGE"

Reprinted From
Road & Track Magazine

ISBN 185520 3480

Published by Brooklands Books Ltd.
with the kind permission of
Road & Track Magazine.

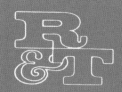

All car enthusiasts love to take to the open road. With any sort of car, new or old, flashy or mundane. It is the pleasure of freedom in virtually as pure a form as we ever experience in modern life. Our friend Peter Egan is the spiritual leader of road trips for all of us who enjoy his writings in the pages of *Road & Track* and our sister publication, *Cycle World.*

In this collection of Peter's *Road & Track* stories about life and adventure on the highways and byways of America, you will find a wide variety of cars serving as the focal point. Everything from a Ford Model A sedan to a Ferrari Dino. One has to admire the courage of any person willing to set off on a long journey at the wheel of a Citröen Deux Cheveaux or an MG TC, but in Peter's case you know that each breakdown or misstep will produce a delightful tale.

So, pull up a chair, make yourself comfortable, and delve into this wonderful collection of Peter Egan stories. And in the words of an old friend, "Every mile a memory!"

Thos. L Bryant
Editor-in-Chief
Road & Track

Our thanks go to John Oakey for allowing us to use his photograph of Peter on our back cover. This was taken in Atlanta in October, 1982, whilst undertaking The Great TC Trek.

Contents

The Great TC Trek

Weaving south on the road to Atlanta

BY PETER EGAN
PHOTOS BY THE AUTHOR

"DRIVING A SPORTS car to the sports car races. Just like the early days," the man said. "I don't envy you guys a bit. I guess that means I'm getting old."

It was three o'clock in the morning at a place called Foreign Car Specialists in Madison, Wisconsin in late October. Chris Beebe, who owns the shop, and I were rubbing a final coat of Meguiar's Plastic Polish into the ancient yellowed side curtains of a dark green 1948 MG TC. Half an hour of laborious buffing had restored them to virtual translucence. Our intention was to embark at dawn's first light on a proper meandering back-road sports car tour that would take us to Atlanta for the SCCA National Runoffs.

The man who didn't envy us was a friend who'd stopped by on his way home from a party to wish us well and had been roped into fabricating some new wiper blades. "Good luck," he said with a benedictory wave on his way out the door. "I certainly hope you make it."

We certainly did too. Every effort was being made.

Chris had been up for three days and nights installing new tires, rod bearings, timing chain, water pump, wheel bearings and kingpin bushings, while I had just stepped off a long airline flight from LA—one of those that stop in Wichita, Des Moines and Dawson Creek to take on fresh flight attendants—and had immediately launched into a comprehensive lube job and cleanup of the car. We'd been planning this magnificent journey for as long as either of us could remember (about two weeks) and didn't want anything to go wrong. The very concept, however, the germinal notion of a long road trip in this particular MG, was far more than two weeks old.

The first time I ever saw the TC it was nothing but an intriguing shape beneath a tarp in the back of a barn. I was living in Madison then and working as a mechanic at Foreign Car Specialists for Beebe, who is best described as an extraordinary guitar player, artist, gentleman farmer (i.e., he lives on a farm but couldn't raise dandelions without professional help), engine builder, mechanic of great finesse, Lotus Super Seven driver and fierce competitor in D Production. He drove at the SCCA Runoffs in 1978 and 1979, finishing 14th and 4th, respectively.

Chris invited a group of us out to his farm for dinner one evening and was showing us around the place when I spied something with wire wheels in the corner of his barn.

"My brother Joe's TC," Chris explained. "He lives on a houseboat in Florida, so I'm storing the car for him."

We lifted the tarp to reveal a nicely restored BRG (almost black) car with black leather upholstery, right-hand drive, bald tires and a set of Brooklands windscreens behind the folded-down windshield. We pumped up the tires and rolled the TC out onto the lawn.

In the late evening light the car was beautiful as only a well detailed TC can be: a fanciful gathering of wood, leather, spokes, green paint, instruments the size of dinner plates, chrome wing nuts, brass, headlights and decorative octagons, all artfully arranged into the archetypal British roadster. Sitting there on the lawn with its low-slung doors, small windscreens and dual-cowl instrument panel, it looked half sports car and half antique aircraft; a Sopwith or Gypsy Moth for the winding road.

As I admired the car's fragile and jaunty good looks from various angles, it slowly dawned on me that I'd never driven a TC, nor even ridden in one. The TC was already a bit of an antique, and a valuable one at that, by the time I reached car-purchasing age. In 1964 you could buy a good (okay, deceptively handsome) TR-3 for $450, but a decent TC cost $1000 or more. I tried to buy a nice red one from a used car lot for $1100 that year, but my efforts ended in failure. The loan officer could find no such car in his Blue Book, my credit status established me as a non-person in the eyes of the bank, and my parents said if I bought the car they'd kill me. I bought the TR-3 instead.

I had raced and/or owned a half-dozen sports cars since then, but had never so much as sat in what was arguably the most important sports car ever to land on American shores. The TC. The Car that Started it All. The Sports Car America Loved First. I noted the oversight, stored the existence of this car away in a mental must-drive-someday file, and waited.

Last autumn the waiting ended. I placed a phone call to Chris.

"Are you going to the Runoffs?" I asked.

"No," he said. "When my engine blew up at Brainerd I lost the point I needed. If I go at all, it'll be as a spectator."

I allowed a brief pause here for silent commiseration, then cleared my throat.

"How would you like to drive a sports car to Road Atlanta this year?" I asked.

"What sports car?"

"The MG TC you have stored in your barn."

The long distance lines crackled and I could hear the wheels of thought turning some 2000 miles away. "Sounds interesting," Chris said, "but I'll have to ask Joe. I think he might like to see the car on the road again. Maybe he could meet us at Road Atlanta. The car will need tires."

"I'll see what the magazine can do."

Joe kindly agreed to let us drive the car, new 4.50 x 19 Dunlops were ordered and shipped, and plane reservations were made. Two weeks later it was three o'clock in the morning at Foreign Car Specialists and the car was nearly ready. We put a final coat of wax on the TC, disconnected the battery charger and locked both sides of the piano-hinged bonnet. At the last minute the shop phone rang and it was our friends John Oakey and Bruce Livermore asking if we wanted a sweep car to follow us down with spare parts. Borrowing a line from Burt Lancaster in *Trapeze*, I said, "No nets." The garage door opened and the TC rolled out into a cold October night.

The first leg of the journey was an 18-mile jaunt to Chris's farm so we could get a few hours of sleep before sunrise. For the sake of purity we drove with the top and windshield folded down. About halfway to the farm Chris noted the night was "a little brisk," which is sort of like calling a Bessemer furnace cozy. Except for the ferocious cold, the drive was wonderful. The engine sounded strong and there were several thousand stars over the open car. I noticed as we drove that the TC occasionally darted and tried to change lanes of its own accord and that Chris was doing a fair amount of steering to keep it straight. "It was much worse before I fixed the kingpins," he said, "but it's still pretty twitchy. I should have put more shims in the spring-loaded tie rod ends."

The only other small worry was oil pressure, which read only 20 psi on the highway instead of a normal 40–50. We pulled over and the pressure stayed the same at idle. Chris tapped the instrument face with his finger. "It must be the gauge, otherwise the pressure would drop at idle." He shrugged and put it in gear. "We'll know soon enough." We made it to the farm without incident.

We were up and on our way south at the crack of noon, driving relentlessly for almost 10 minutes before stopping for lunch in Evansville at a main street cafe. Getting out of the car was not easy, as we'd packed ourselves in. The TC has a very small luggage space behind the rear seat, so once that's filled everything else goes in the footwell or gets strapped to the spare tire on the back of the car. The TC allows about the same amount of personal luggage as a motorcycle, but you can put it to better use. We used one sleeping bag to prop up the driver's left leg for better thigh support, jammed one of Chris' rubber boots between the clutch pedal and transmission tunnel to make a footrest, and wedged a leather covered pillow (a legacy of Joe's touring days) between the driver's right knee and the protruding inside door latch. With all that impedimenta in place, the driver's seat is remarkably comfortable, but you think twice before jumping out of the car if someone shouts fire. The passenger seat is comfortable with no modifications other than a lap blanket, to make up for the lack of heater.

At the cafe in Evansville we got out the road atlas to find out where we were going, something neither of us had done before this moment. Chris's road atlas, it turned out, was a very early edition with many of the modern roads missing, apparently printed right around the time of the last Sioux uprising. Illinois looked very long on the map, occupying two full-length pages. We both glanced out the restaurant window at the TC and then looked back at the map of Illinois. "Maybe we could take a great circle route over the pole," Chris suggested, "and avoid Illinois."

We decided to take Illinois Highway 1 all the way through the state and cross into Kentucky by ferry boat on the Ohio River. We packed ourselves back into the car and headed south, through the industrial brick of Rockford and into the small towns of northern Illinois.

It was one of those beautifully clear Midwestern autumn afternoons when the sun is warm but the air feels like bracing aftershave and seems to have come down from a Canadian jet stream straight into your lungs. And with the top down on a TC that's exactly what it does. The car was giving off a heady combination of aromas, a mixture of old stiff leather, mouse leavings, cotton wiring loom and hot oil chilled and reheated. It smelled almost exactly like an old Aeronca Chief in which I once took flying lessons. Being on the ground, however, the MG was fractionally safer.

When Chris gave up the wheel to me, I discovered he'd been doing a remarkable job of staying one step ahead of the twitchy steering. It took concentration and planning to keep the car in its own lane. Otherwise, the TC had a pleasantly hard, mechanical quality to the controls; the feeling of being a piece of machinery first and a car second, as though, without wheels, it might be equally happy as a printing press or a metal lathe. The raspy exhaust note matched its mechanical essence perfectly, the engine sounding bigger and meaner than its 1250 cc, more akin to a TR-3 or something similarly large.

Our car was louder than I remembered them being in stock form, and Chris attributed this added snarl to some bygone baffling in the muffler. The engine was also a high revver for its era (at least by American standards), cruising at about 4200 rpm at 60 mph. We hit 75 mph passing a diesel car (had to; the exhaust was making us sick) and the effect was a magnificent live-from-the-Mulsanne Straight wail. In 1948 sports cars sounded nothing like sedans.

Our first gas stop yielded a slightly disappointing 21 mpg, a figure Chris remembered as being about average for TCs. The oil was down one quart, which we had expected, given the volume of smoke that enshrouded us at every stop sign. The rings were a little tired. A close inspection of the engine revealed that absolutely nothing was seeping, dripping or leaking from any joint or junction. Chris had taken great pains with gaskets and silicone sealer when assembling the outer engine pieces, and

everything under the hood was spotlessly clean. Mileage and oil consumption remained fairly consistent throughout the trip.

The small towns of the Midwest were in the grip of Halloween and Homecoming, full of color and kinesis, like some giant Brueghel tableau of vibrant village life. From the car we could see high school girls with cheerleader jackets and pompons, schoolbuses with flashing red lights, brilliant yellow maple trees, hardware store windows filled with deerhunter red and shotshell displays, leaf raking and storm window washing, all surrounded by yellow corn in fields at the edge of town.

Then we entered the larger town of Kankakee to find a motel for the night and everything changed. Parts of the inner city were old and lovely, but typically the outskirts were a moat of everything in the world that is quick, cheap or painless. Out of town, on the outskirts, near the bypass, off the Interstate, it was the home of square hamburgers, clean beds, fiberglass shower stalls, tacos in Styrofoam coffins, English Olde Pubs with particleboard rafters, jaded maids and a Steak & Spirits. Worse yet, Libations. We stayed in a chain motel (Welcome Shriners), had something microwaved for breakfast and took off down Highway 1, vowing to stay in small town main street hotels or camp for the rest of the trip. The TC was hopelessly out of place within 10 miles of any Interstate exit ramp.

A great rain began falling at mid-morning so we stopped to put up the weather gear on the TC, which is sort of like dressing a child to play in the snow. It takes a little time and care. The side curtains are stored in an elaborately conceived wood and canvas pouch in the rear of the boot, each of the four panels separated from the others by a protective flap, all held together by a wooden wedge that appears to have been whittled by someone's uncle from rural Yorkshire.

There is no heater in the TC, but with the side curtains up the car is relatively snug. The floorboards and transmission tunnel give off enough heat to keep cockpit temperature comfortable in chilly autumn weather, but survivors report this system falls short in the dead of winter.

The rain turned into a downpour and water began to blow in around the top and bottom of the windshield, raindrops spitting and sniping at us from odd angles like loose rivets in a boiler room. We dammed the bottom with a chamois and stopped at a hardware store in Watseka, Illinois to buy some rope caulk for the top. As we laid a bead along the top of the windshield, Chris said, "The hardware store is the TC owner's friend."

"We'll laugh at the weather now," I said, slipping behind the wheel. The interior stayed dry, and I was amazed at how well the overlapping side curtains and top kept the rain out. The wipers were run by an overhead motor bolted to the top of the windshield and linked by a little chromed Rube Goldberg arm. Chris noted the wiper blades were hypnotic and moved like a pair of skaters with their hands behind their backs. "Sounds like it's time to stop for coffee," I said.

We stopped at a roadside cafe where our place mats were tourist maps of the U.S., with each state represented by a symbol. Wisconsin had a cow grinning from behind a large chunk of Swiss cheese and Georgia appeared as a huge peanut. I pointed out our route to Chris. "Okay, we start here at the cheese, drive over Abe Lincoln's hat, head down through Daniel Boone's fort, across the flank of this thoroughbred horse and straight into the big peanut." The place mat was better detailed than our old road atlas, so we took a copy with us.

In late afternoon the weather warmed and cleared, and at the lower end of Illinois we began to see homes in the southern style, with low slanted roofs sloping down over welcoming front porches set out with chairs and porch swings. The more vertical, Teutonic architecture of the North became less common. The TC was still holding 20-psi oil pressure and sounding fine, though it smoked a great deal every time we stopped and idled.

We drove very carefully in urban traffic because the TC's brakes (or lack thereof) created some heart-stopping moments when cars pulled out in front of us. Chris read a passage to me from the original owner's manual that warned not to step on the pedal too hard because "the MG TC brakes are extremely smooth and powerful." The man who wrote that may have been using the clutch pedal for comparison. The TC brakes didn't so much stop the car as throw a wet blanket over its long-term progress.

As anyone who has ever driven through Illinois can tell you, its roads are among the worst in the country, with expansion strips about every 30 ft and plenty of potholes to break the monotony. The TC did not actually drive through Illinois, but passed over its roads in a series of small leaps and frisky hops. South of Danville we saw an amusing road sign that said BUMP, which, as warnings go, is about like having a sign that says BEWARE OF THE CORN in Nebraska.

We arrived at the Ohio River too late in the day to take the ferry across to Kentucky, so we decided to camp at a state park overlooking the river and eat at a cafe in the town of Cave In Rock. We had a dinner of batter-fried catfish, French fries and deep-fried corn fritters. We considered asking the waitress for some fried coffee, but she was a big, powerful woman and we thought better of it. The crunch of fried batter in the restaurant was drowned out only by the electronic beeping of six video games operated by teenagers. "I gotta put up with that racket all day," said our waitress, "but at least there's more money in it than there is in serving food."

As we climbed back into the MG, Chris said, "I wish this were a time machine."

The starter drive engaged with its customary metallic clang, and I said, "Maybe it is."

We erected our tent on the bluffs overlooking the river by the parking lights on our faithful TC and spent the night holding down the tent corners when a tremendous, near-biblical wind arose out of nowhere. We were under a large tree and something (acorns?) kept dropping on our tent all night. In the morning we were awakened by a woman's voice asking if we were awake. Denials were useless, so we said yes. Peeking out of the tent we saw a woman wearing horn-rimmed glasses, a large hairdo and a ranger's jacket. She was accompanied by a dog with three legs and an old man with a brown paper bag.

"Y'all owe two dollars for this campsite," the woman said, checking her clipboard. We got up and paid.

The old man, a heavyset fellow in overalls and a DeKalb Seed cap, was stooping to pick things from the ground and put them in his paper bag. We asked what he was picking.

"Kongs."

"Kongs?"

"Yep."

"What are kongs?"

He showed us a brown nut. "Them are pee-kongs," he explained.

"Ha!" Chris said to me, "Pecans."

"That your MG?" asked the old man.

"Yes."

"Reminds me of a airplane my son took me up in after the war. Doors didn't even lock. I didn't think much of that." He went back to work and Chris and I munched on a few kongs as we broke camp.

Throughout the trip I was amazed at how many people, young and old, identified our car as an MG without giving it a second glance. Occasionally someone on the main street of a small town would even say, "Is that your TC?" knowing it was not a TD or a TF. Waitresses, gas station attendants, everyone seemed to know someone who now had one or used to have one. Maybe all nice old roadsters with wire wheels and clamshell fenders are MGs to most people, and if we'd had a Morgan, that would have been an MG too. In any case, the look of the T-Series MGs made a big impression on middle America a long time ago, and few people have forgotten. Happily, no one ever looked at the car and asked if it were a replica.

We drove down to the river and onto the car ferry, crossing the Ohio for $2.00. It was a fine clear morning with whitecaps on the river. We took the top down while we were on the ferry, but left the side curtains up. This gave us the feel of open-air motoring without as much cold blast on the back of the neck, a good arrangement for sunny but brisk autumn mornings.

When you enter Kentucky from Illinois, the countryside is immediately more wooded and hilly. The stereotype of the Old Kentucky Home is well founded, as the hills are full of secluded little farms and cabin-like homes with low-sloping roofs, broad porches and trails of white wood smoke rising from their stone chimneys. The distinction between Illinois and Kentucky is well marked by both architecture and geography. You get the feeling that all of the early settlers who liked farming and community life settled in Illinois, while those with a taste for squirrel hunting and bourbon moved to Kentucky. The upshot is that a more romantic aura hangs over Kentucky. You have Kentucky bluegrass, Kentucky Derby, My Old Kentucky Home, Kentucky Fried Chicken, Kentucky windage, Kentucky Long Rifle, etc. "Is there an Illinois anything?" I asked Chris.

"The Illinois Tollway."

We took a beautifully winding road into Marion, Kentucky, a pretty town with a courthouse square and a restaurant sign that said, "Coffee Shop, Home of Biscuits and Gravy." You have to respect a sign painter who has your number, so we stopped for breakfast. I'd wicked up a lifetime supply of damp and clammy while camping the night before and needed coffee. The menu included grits, country fresh ham (none of this city stale stuff) and biscuits and gravy. We were, as the waitress told us, "a long way from West Consin." (Many people in the South call Wisconsin "West Consin," as though there might be an East Consin.)

As we rolled south into the country I realized we'd both picked up a new habit; we found it impossible not to wave at motorcyclists and farmers driving tractors or harvesting equipment. We seemed to have more in common with them than with other motorists; a kinship of wind and open air.

As with an old horse and buggy, there is something about the MG that suggests you are not walking, but riding; seeing the country from a self-propelled vehicle rather than by your own physical effort. Maybe it's the likelihood of having to walk that makes it more fun to ride. A more modern car separates you from the scenery and weather by its technical proficiency. It simply owes you transportation and you expect it. The TC doesn't owe you anything. That you've come this far—no matter where it is—is a minor miracle, so the sound of a healthy engine makes you happy. Nieuport pilots probably felt the same about their LeRhone radials over enemy trenches. When the sun is shining on the hood and the engine is running smoothly, you keep thinking it's good to be alive.

Just out of Marion, Chris pulled over at an abandoned roadside shack to check the oil, so I wandered around to take rustic car/shack pictures. We were getting ready to go when a patrol car suddenly careened around the corner and skidded to a lockup stop in the gravel behind the TC. A short, heavily built cop got out and walked up with his hands on his hips. He was ready for whatever we had. "What you boys doin' here?" he asked.

"Adding a quart of oil and taking some pictures of this car."

He looked at us, one at a time. Then he looked at the car. He looked at our shoes. He weighed everything he was seeing. Chris was wearing a red beret and I wore a stroker cap. The red beret didn't look good to the cop. You could tell that. He pointed up the hill to a farmhouse. "That old boy called me. He's a little worried about you boys here. Says you been stopped for a good 15 minutes."

"Well, we're just adding a quart of oil and taking some pictures of this car."

He looked at the empty quart of oil and the car. "It's okay, I guess. Have a good trip." He tipped his hat, climbed back into the patrol car and sped away. We had been duly visited and checked out. We were okay, even if we had funny hats and a funny car. Not criminals. Just officially odd.

We drove down to Hopkinsville and skirted the Ft Campbell Military Reservation, something I would love to have done in 1969—a very good year for cleaning rifles and crawling in the mud—but couldn't. We then took State Highway 100 halfway across south Kentucky. This stretch of road is now permanently traced in my own road atlas with the legend Sports Car Road scrawled beneath it.

Highway 100 is like something from a Woodsey Owl cartoon, a smooth ribbon of dipping and weaving road over an idyllic rural landscape of tobacco barns, woods, cabins, farmhouses and small junctions that pass for towns. The MG, despite its twitchy steering, was a joy to drive on this road. It occurred to me that the TC was not exactly the fragile, slow little antique I had imagined it to be at the beginning of the trip. We cruised at 60–65 mph, passed other cars with ease at that speed, and progressed across the map quickly. The skinny tires provided surprisingly good roadholding with easily controlled drifting and sliding at the limit, and the coal cart suspension kept body roll to an absolute minimum.

To drag out the old sports car mythology, a TC really does go around corners as if on rails, making up with high average speed what it lacks in brakes and sheer acceleration. It was amusing to see big American cars, and even a few small sporty sedans, wallow and screech around corners as they tried to keep up with—or ahead of—the rock-steady and unflustered TC. It doesn't take much imagination to see why the MG won the hearts of drivers who stepped out of their American sedans in 1948.

The engine sounded a little clattery when we started it after lunch, so we decided to experiment by adding a can of STP. Our oil pressure shot up from 20 psi on the highway to nearly 22 psi. Chris took a pencil and did some quick calculation. "If we add 15 more cans of STP to the crankcase, we'll have normal oil pressure."

Late in the day we debated whether to camp or find a motel. I considered our previous windy night under the kong tree and cast my vote. "Let's pitch a room and sleep in a motel under the stars."

We stopped for the night at a mountaintop lodge called the Alpine Hotel overlooking Burkesville, Kentucky and watched the Brewers get trounced in the last game of the World Series. A sad day for the state with the grinning cow and the big cheese.

In the morning a thick fog surrounded the mountaintop, and I was glad to see it. Between the sunburn and windburn from two days of top-down touring with the TC, I looked like I'd been shaving with a blowtorch. A darker sky was welcome.

We couldn't see much of the surrounding countryside because of the fog, but it didn't matter because the TC is a constant entertainment in itself. If the scenery is no good, you just look down that long, narrow hood and watch the reflection of the road and sky go past in the chromed headlight housings. That, or look at the polished wingnuts, windshield support castings, instruments, leather upholstery and steering wheel—you look *around* yourself—and wonder if the craftsmen who made those pieces are still alive. Could anyone duplicate them now, especially at the price of an economy sedan? The TC cost less than $2000 when it was new. Descended from the TA and the TB of the Great Depression, it was built to be fun and cheap, yet had a depth of quality that made it perpetually satisfying to contemplate as you traveled. Odd how the wood grain of a dash and the right sweep of a fender line could be so much more endearing to the human spirit than a digital speedometer or a recorded voice telling you to shut the door.

If you had to drive everywhere all your life in cars like the TC, you'd rebel and insist that engineers come up with something quieter, drier, stronger and easier to get in and out of, but as an occasional experience, the TC made a wonderful antidote to the cautious sameness and sensory void provided by many newer cars. It was a rolling reminder of what we've given up to gain comfort and convenience.

A Crazy Somebody's Fireworks sign told us the Tennessee border was near, and we crossed over on Highway 127 just as the fog began to burn off. Tennessee. Home of the Tennessee Waltz, Tennessee walking horses, Tennessee sour mash bourbon, Tennessee mountain home, Tennessee hillbillies and Tennessee Ernie Ford, not to mention the Grand Ole Opry, the Nashville sound, the Chattanooga Choo-Choo, the Gretsch Tennessean electric guitar, Chuck Berry's Memphis, W.C. Handy, Beale St and Elvis. Amazing what pretty hills, music and good whiskey will do for a state's character. "Name something from Ohio," I told Chris.

"The Ohio Turnpike."

Tennessee is a state of parallel ridges where the main roads go down the northeast-to-southwest valleys, and the other roads that link them are wild and curving. TC country again. When you drive along the Tennessee River Valley, it's easy to see why the TVA had so much trouble moving people out to make room for their dams and man-made lakes.

Red clay and tall pines marked another natural border as we entered Georgia. The TC hummed through the hills of the Chattahoochee National Forest, and by nightfall we were idling into our annual favorite campsite at lovely Lake Lanier. The lake is about a 20-minute drive from the Road Atlanta track, and the shoreline was lighted by the campfires of racing people from all over the country. We put up our tent by headlight and wandered over to warm ourselves at the campfire of Bob Shadel (F Production Midget driver) and friends. We drank wine, warmed our hands and talked racing until late at night. It was growing very cold and no one wanted to leave the fire for chilly tents back in the shadows.

We were finally too tired to stand up any longer and returned to our tent. I have a down sleeping bag I got on sale for $26 at a sporting goods store several years ago. It isn't the same bag Hillary used on Everest, but it may have been standard issue on the ill-fated 1972 Taiwanese Expedition in which everyone froze to death. The bag has about nine feathers in it, and by morning I was lying on about seven of them, fully dressed with my teeth chattering. Luckily, the campground had hot showers, and at 5:30 a.m. I took one that lasted until about eight o'clock.

After searching, we found Chris's brother Joe and his friend Maria, as well as our non-sweep drivers John Oakey and Bruce Livermore, whose Aspen wagon had chewed up an axle bearing on the way down. We cautioned them to take a good reliable British roadster next time, preferably one from the late Forties.

Joe was quiet and thoughtful as he examined his old car, probing here and there and running his hand lightly over the smooth lacquered surfaces he'd sanded and painted in an earlier part of his life. "Warm Black," he said. "I mixed the color myself. Black with a little bit of green in it." Joe took the car for a drive around the campground, shifting and clutching with the same effortless smoothness as Chris. Their father, Chris told me, taught the whole family to drive without the clutch so they would learn to match revs perfectly.

Joe told me later he'd bought the car for $1000—representing everything he could sell, beg and scrape together—from a Denver woman in 1962. The car was white then. Someone ran a red light and nearly destroyed the car the following year, so he took it down to the frame and spent three years of nights and weekends on its restoration. After that it got a lot of use, crossing the country several times.

I looked at the TC and wondered if there were any cars made now to engender that kind of zeal and enthusiasm, to cause a young man of no means to sell everything he owned just to have one, then to work three years to make it like new after a disastrous accident. In the year 2016 would I fly two thousand miles for a chance to spend two weeks in a 1982 car because of its sparkling character, good looks and historical significance? If the Eighties answer to the TC was out there somewhere, I hadn't noticed it.

The MG spent the next three days parked in the paddock at Road Atlanta while we wandered around soaking up the ambiance of the Runoffs and watching races. Every time we strolled past the car there was a small crowd around it, taking pictures or just looking.

After the last race on Sunday evening we returned to the TC and found a man and his wife down on their knees inspecting the front suspension in the near darkness. As we approached, the man let out a loud whoop of joy. "It's not VW! It's real!" he shouted to a band of passers-by. "God bless this car, it's real! It's not a replica!"

It took us five days to get home, meandering through the South on winding secondary roads, over narrow bridges, through little towns, among the Blue Ridge mountains and the hills of Tennessee and Kentucky. The TC, with its nearly 22 psi of indicated oil pressure, soldiered on without miss or malfunction. Even the Lucas electrics, fuel pump included, continued to work flawlessly.

We backtracked through autumn, watching the brilliant yellow and red hardwoods of the South fade to a more uniform

brown as the latitude changed. Somewhere near the Wisconsin border a cold wind picked up and the leaves blew horizontally across the road without touching it, like an angry warning of something less friendly on the way. The clouds were suddenly lower, darker and faster moving than before. Snow flurries appeared. It was a good day to be ending a trip in an MG TC with no heater.

The Wisconsin Capitol dome appeared over a wooded hill, we merged into city traffic and 10 minutes later we were home. Someone opened the electric overhead door for us and we drove into Foreign Car Specialists just before noon on a Friday. We'd been gone 12 days and the Jaeger odometer in the TC's dash said we'd traveled 2600 miles, for a total of 88,080 miles on the car. One hundred and ten gallons of gasoline had been consumed, along with 11 quarts of oil and two cans of STP.

Chris hesitated before shutting off the engine, his hand poised on the key. We looked at the instruments. The oil pressure was stable, the amp needle hovered slightly to the plus side of center, the idle was even and the engine sounded as healthy as it had the night we left. He turned the key and it stopped running.

We sat motionless in the car for a moment, not wanting to get out. I laid a gloved hand on the dash and patted it lightly. "This is a good old tough car," I said.

Chris stared down the bonnet and whispered the word, "Nothing," with wonderment in his voice. "Twenty-six hundred miles. Not a leak, a miss, a burned-out lightbulb. Nothing. I'd have expected less from my Chevy van."

Chris later admitted he'd measured the TC before our trip to make sure it would fit on his Lotus Seven trailer, "just in case." I was then forced to confess that I had about 2000 miles worth of bus fare tucked away in a hidden corner of my billfold. We were not superstitious, but these things were discussed well out of hearing range of the TC, and only after the trip was safely over, knowing that the crystalline structure of metal in 34-year-old cars is highly sensitive to the power of doubt.

That night we went out with friends for a celebratory dinner during which we toasted the South, the North, the Runoffs, West Consin, self-envy, eternal youth, Abingdon-on-Thames, Morris Garages, Lucas fuel pumps, brother Joe and, of course, the TC.

On the way home from the restaurant later, the highway traffic made me a little nervous. I wasn't certain if I'd already paid for the TC trip or if I'd have to settle up at some future time, and didn't want the MG involved in the inevitable scrape with cosmic justice. Halfway to the farm I decided to write the trip off against the two years I spent in the army, and accelerated into the night in a cloud of smoke and a great wail of vintage sound and fury. The TC was a real sports car, after all, and cautious driving was nothing less than a breach of faith.

THE CORVETTE RACER

A Sprite driver's view, from a safe distance

BY PETER EGAN

DRAWINGS BY LEO BESTGEN

To those of us who raced our diminutive H Production, Sprites and similarly high-pitched cars in Midwest Council and SCCA regionals, Corvette drivers were always a breed apart. We spent many weekends pitted next to Corvette teams, adjusting our pathetic tiny Austin valves or performing some other jeweler's chore on our Bugeyes, watching these fiberglass behemoths and the crews who serviced them out of the corner of one collective eye (we didn't have the nerve to stare with both eyes). Corvette drivers and Sprite drivers didn't mix much, except to borrow the occasional strand of safety wire or half roll of duct tape from one another, but a shortage of information and personal contact didn't prevent us from forming some well developed stereotypes and generalizations.

We knew a few things about Corvette guys:

Corvettes always seemed to be raced by drivers who had crew cuts long after everyone else in the club looked like George Harrison. You got the feeling that if they weren't driving Corvettes, they'd be dropping you for 50 pushups at Camp Lejeune or chewing Copenhagen without taking it out of the can. Corvette guys drove in black Wellington boots while everyone else wore effete moccasins or Nomex booties. They wore open-face helmets with no visors so everyone could see them scowl. When they dressed up to go out at night they wore white nylon windbreakers, white pants, Goodyear caps with the visors pulled down over the eyes and at least one article of clothing with stars & stripes or

crossed checkered flags on it, and they always went out for pizza and beer in a place with bright neon lights.

Corvette drivers adjusted their points with air wrenches and their pit men were sometimes crushed to death by fallen lug nuts. Corvette guys had trailers with six wheels and towed them with brand-new extended cab white pickup trucks with more checkered flags and stars & stripes on them. The trailers had overhead tire racks with tires that were exactly four times as wide as they were tall. These tires cost $600 a piece and lasted only one practice session before being torn to bits by raw torque.

Corvette drivers never asked any questions at drivers' meetings.

When a Corvette driver started his engine the smoke and noise came out of side pipes the size of storm sewers, and the exhaust pulses threw large chunks of gravel and blew your tent down. When these engines blew up on the track, the concussion and coolant spray left corner workers dripping and dazed for hours. If a Corvette threw a rod through its sump, the resulting oil spill made your Sprite go backwards through Turn 3 for the rest of the season.

Every Corvette driver had at least one gold tooth. Corvette drivers made their money in cattle or lumber, or else they owned more than one gas station and paid men named Frank to run it for them. Corvette drivers themselves had names like Bart or Chuck or Bob, and their last names were usually Johnson. They always had Texas jet pilot accents even if they were from Michigan's Upper Peninsula.

Corvette drivers used hacksaws to set their camber and had tool chests where each tool occupied a whole drawer. The sockets at the small end of their socket sets started at 1¼ in. and went upward to sizes that frightened Caterpillar mechanics. Between races, Corvette drivers arc-welded things to their chassis in a blinding shower of sparks. When a Corvette driver jacked up his car, he didn't so much lift the car, as push the earth away from it.

Male Corvette drivers had blonde wives who chain-smoked and had cattle ranch tans and pale blue eyes, while women Corvette drivers were always single because they couldn't find anyone who was man enough to marry them. Corvette drivers never lit their cigars. They just chewed them flat and walked around the pits until they saw the front suspension on a Lotus Seven or the engine in a Sprite. Then they threw the flat, wet cigars on the ground in disgust.

Corvette drivers used approximately 100 gallons of Union 76 racing gas on every lap and had government surplus fuel cells from armored vehicles. While those of us in Sprites, Midgets and Spitfires had to *drive* from one end of the main straight to another, Corvette drivers simply *launched* themselves in a great belch of power and landed at the other end on four smoking tires. Corvette drivers drove on a much shorter track than we did, and their pitboards flew by like fastballs and were impossible for the human eye to read, while we had time to examine the small print on the race marshal's badge-littered vest.

When Corvette drivers massed on the starting grid for the A and B Production races at Elkhart Lake, the announcer used to say, "Ladies and Gentlemen, it's time to shake the dew off the lilies." He didn't say that before the other races, because nothing else thundered, rumbled and shook the ground quite the same way.

Not even Sprites.

11

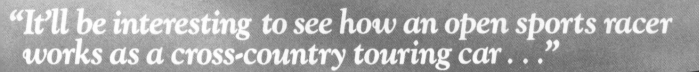

"It'll be interesting to see how an open sports racer works as a cross-country touring car . . ."

NORTHEAST BY WESTFIELD

"What year Undo is that, mister?"

BY PETER EGAN

PHOTOS BY THE AUTHOR, BARBARA EGAN, CHRIS BEEBE &
SOME GUY WHO WAS JUST STANDING AROUND

SOUTH OF AMBOY, California, the Mojave Desert is not really sports car country. It looks like a good place for an A-bomb test or a Sabre-Jet attack on a giant mutant spider. It is mirage country, where everything shimmers in the distance, and you cross the white and dazzling salt flats of Bristol Dry Lake half expecting T.E. Lawrence to come riding out of the heat waves. The roads through this territory run straight as the crow flies, only it's too hot for crows and there's nothing for them to drink so there aren't any.

There aren't many open British sports cars, either. In fact we were nearly alone on the highway, which on an afternoon in early June was positively humming with heat. My wife Barbara and I were on our way to Elkhart Lake, Wisconsin. We were cruising across the Great American Desert in a small English car designed a quarter of a century ago to run on narrow green lanes and mist-shrouded race tracks in a country where sunlight is about as common as a lunar eclipse. I was behind the wheel, watching the water temperature gauge with one eye and trying to remember if there was any place in England where you could drive upgrade all day in 110-degree weather. I didn't think there was.

This particular English sports car was a Westfield. The Westfield is a replica of a Lotus Eleven, a very successful sports racing car from the late Fifties and one of the cars that made Colin Chapman a household name in households that love fast, agile, nice-looking cars. Just three weeks prior to our trip, the Westfield had arrived on *Road & Track's* doorstep as a disassembled, unpainted kit. Driven by some kind of tangential homing instinct, I'd spent all of my nights and weekends virtually living in the R&T garage, subsisting on a health fad diet of Coke and cigarettes, trying to assemble the car in time to make the June Sprints.

No Midwesterner, fallen-away or otherwise, likes to miss the Sprints, which is an annual rites of spring tradition held at what may be the most beautiful sports car track in the country. The event is half SCCA National and half medieval festival, drawing a huge number of entrants from all over the country. An added incentive, if we needed any, was that our friends Chris Beebe and crew would be racing his D-Production Lotus Super Seven.

The night before we left on this grand trip I was still working on the car (proof that work expands to fill the time allotted) when our Editor Emeritus, Tony Hogg, stepped into the lighted chaos of the R&T garage. Tony was especially interested in the project because he had owned and raced an Eleven in England and Europe during the late Fifties. Tony lit a cigarette and asked how things were going. Pieces of the car were still scattered all over the floor.

"Almost finished," I said. "The car has to be ready to go tonight because Barb and I are leaving for Elkhart in the morning."

Tony raised one incredulous eyebrow. "You're driving that bloody thing all the way to Elkhart Lake?"

"Yes," I said. "It's going to be sort of a motorcycle trip on four wheels. The Westfield doesn't have a full windscreen and the convertible top isn't in production yet, so we're taking rainsuits and helmets. It'll be interesting," I added, "to see how an open sports racer works as a cross-country touring car."

Tony gazed at the Westfield for a few moments and took a thoughtful drag on his cigarette. "I should think," he said in his best detached British way, "that it'll be just bloody awful."

Awful, of course, was a matter of perspective. Barb and I had crossed the country a half-dozen times on various motorcycles, so the prospect of leaning back in a comfortable seat, largely protected from the wind, seemed relatively luxurious. The Westfield also had more luggage space than most motorcycles. There was storage space on the wide armrest panels inside the doors and on the floor beneath the arched knees of driver and passenger. The rear body section swung upward, revealing a flat tray that was part of the floor pan. This area was subject to dust and water spray from the rear tires, but luggage could safely be wrapped in plastic garbage bags and lashed to the frame and spare tire. The Westfield was not the Globemaster of cars, but we wouldn't have to leave our toothbrushes behind.

The only slightly awful part was taking off on a 5000-mile journey with an untested car, with virtually no time to run the engine in. Adjustments would be made on the road. I packed enough tools to field-strip the car if necessary and repair nearly anything en route. Fortunately the Westfield is a very simple car

"We pulled out of our driveway into a hot summer morning; mad dogs and English car, off toward the noonday sun."

held together with only a few conventional sizes of nuts, bolts and other fasteners, so the toolbox was small.

I finished the car at midnight, drove it back to our house and packed what few clothes I needed. I told Barb we should get up at 4:00 a.m. and leave in the early morning darkness to avoid crossing the Mojave in the afternoon heat, so we set our alarm and went to bed.

Getting up at 4:00 a.m. sounded like a much better idea at midnight than it did when the alarm went off at 4:00 a.m. We went back to sleep and eventually got up at 9:00 a.m. By 10:30 the car was packed and we pulled out of the driveway into a clear, hot summer morning; mad dogs and English car, off toward the noonday sun.

Part of the plan on this trip was to avoid Interstates at all cost, so we crossed the California desert part way on Highway 62 and then turned north toward Amboy. This is all 2-lane blacktop that runs through such oases as Yucca Valley, Twentynine Palms and Essex, with miles of cactus, Joshua trees and empty desert between.

By the time the freeways of greater and lesser L.A. had spilled us onto these emptier roads, it occurred to me the Westfield had a remarkably good highway ride for a taut, light sports car. Not nearly as jittery as might be expected.

The added weight of our luggage no doubt helped, and the car cruised serenely at 60 to 70 mph with a sort of hunkered-down bantamweight Cadillac feel. A bit too hunkered down at times. Even with the front springs jacked up all the way on their platforms, the low-slung oil pan occasionally scuffed the pavement and random dead things.

Otherwise, the Westfield made a fine highway car. The side-mounted BSA muffler gave out a lovely, throaty purr. Even at 75 mph the cockpit was relatively tranquil. (Don't ask relative to what. An N2S Stearman, maybe.) There was some backdraft on the passenger's neck, but the headrest/fin did a remarkable job on calming the air around the driver's head. In normal driving position you looked over the windscreen rather than through it, but the plexiglass deflected nearly all of the wind over your head. Stones and insects, however, refused to follow the wind flow, so goggles or a helmet with a face shield was nice, especially around gravel trucks.

Until about one o'clock in the afternoon the car ran perfectly, with the water temperature needle centered on Normal. After that the desert temperature climbed to just over 100 degrees and the needle moved halfway to Hot. When we started a long upgrade climb near Yucca Valley the needle pegged itself solidly on Hot. The intense engine heat, aided by an exhaust header that wraps around the front of the left footwell, also began to warm up the driver's compartment. The pedals got hot and the handbrake lever began to burn my leg, so I wrapped a sweatshirt around the handle. We stopped at a country store, parked under a tree and got out to cool down for a while.

I poured down a quart of Gatorade and raised the Westfield's hood to look things over. I smelled gasoline and discovered that the exhaust header heat was boiling fuel out of the float bowl on the rear SU carb. Willard Howe, the U.S. Westfield importer, had warned me this could happen on blistering hot days, so I had cut a vent in the aluminum bodywork behind the carb to let the hot air out. This, obviously, was still not enough. The Westfield was a brand-new car, untested in this climate and being driven on a true shakedown run, so I began a list of small improvements and modifications to be made when we got to Chris Beebe's shop in Wisconsin. I got out a notebook and wrote: "1. Asbestos heat shield for driver's footwell. 2. Improve heat shield between carbs and header. 3. Make cold air duct for carbs. 4. Move radiator into nose with full shrouding and fit electric fan."

I poured some cool water on the carbs and we took off, climbing and warming our way over the Bullion Mountains and

NORTHEAST BY WESTFIELD

down onto the flats of Bristol Dry Lake.

In Amboy we came across an inviting gas station with a willow tree and a CAFE sign with red letters 12 ft high. A perfect place for pulling off the desert. The cafe was cool, bright and polished inside; lots of stainless steel coolers with that sweating, refrigerated look. We had iced tea, soup and coffee with a side of ice water. "This is a nice place," Barb said. I was nodding when a man with a pressurized canister and nozzle asked us to move our feet so he could spray the crack along the bottom of the counter.

"What are you spraying for?"

"Roaches."

"Talk about service," I said when the man was gone. "You won't find roach spray much fresher than that."

We filled our almost-empty fuel tank and calculated our mileage at 44 mpg. Not bad for a fast car with a new engine. The tank held only 5 gal., but with that mileage we didn't have to start looking for gas stations much before 160 miles. The Westfield's small frontal area and aerodynamic shape certainly didn't hurt its fuel economy; at highway speeds the car felt almost as if it were traveling through a vacuum, and its ease of acceleration at passing speed added to the sensation. With the big MGB wire wheels and the standard 4.20:1 Sprite rear end, the engine was turning a fairly relaxed 3700 rpm at 60 mph.

We passed through Essex, California (pop. 150); one of those places that makes Paris look like a huge glamorous city near a nice cool river. A sign on a building said, "Thanks to Johnny Carson, Essex has TV!" I figured there was a story there somewhere, but it was too hot to stop and ask.

Turning onto the inevitable Interstate long enough to get ourselves out of California, we headed down I-40 toward the green banks of the Colorado River, past the rock spires of Needles.

On the Interstate the Westfield attracted an incredible amount of attention. People rolled down their windows to ask what it was, elderly couples smiled and waved and a Bible school bus listed to one side as all the children ran to the windows. Cars followed, passed, dropped back and followed again to get a better look, and people going the other way on the Interstate actually honked and waved from across the median. Wherever we stopped, the Westfield cre-

ated a minor interrogatory riot.

Nearly everyone asked (1) what kind of car it was, (2) what kind of engine was in there, (3) was the engine in the front or the back, (4) how much did it cost, (5) what kind of mileage did we get, (6) where were we going to and coming from, and (7) what were we going to do when (not if) it rained. A few fully grown people asked if they could just please sit in it for a minute. At a gas station I helped a husky, sunburned 50-year-old farmer in overalls slide down into the passenger seat. He looked around himself and said to his wife, "Now this is all right." I felt as though we were barnstorming the first airplane across the country. I think we could have charged money for rides. The car had some elusive magic of shape and scale that stirred the imagination.

We passed through Kingman and late in the afternoon began climbing into the

cool green Arizona mountains. After a day in the desert the pine country of the Prescott National Forest was like another planet. Just after sundown we swerved to avoid three deer crossing the highway and decided the nearby town of Williams would be a good place to stop for the night. We checked into a motel, covered the car and walked off in search of a stiff drink.

In the morning I got up early to retorque the cylinder head and adjust the valves on our green engine. An Australian gentleman walked out of his motel room and said, "Ha! A Lotus Eleven! With the car cover on in the dark last night I told my wife it was a D-Type Jaguar. Nice car, either way."

We stopped for fuel and the kid at the gas station walked around the car, obviously looking for an emblem or some sort of make identification. He finally got down and read the word engraved in the knock-off hubs, and I saw him mouth the word, "UNDO." I thought for a moment he was going to ask the classic question, "What year Undo is that, mister?" Alas, people are more sophisticated these days, and in the end he asked, in his best Steve Martin imitation, "Now what kind of a dang deal *is* that?"

Barb took the first stint at the wheel

and we headed up to the Grand Canyon in clear, cool morning weather. The car was running beautifully. When we entered the National Park, people seemed to be leaving in droves, but we were almost alone going in. I hypothesized that Editor John Dinkel was holding court at the Main Lodge, telling his favorite puns. We drove to the rim of the Grand Canyon and I tried to stifle my usual pathetic reaction to the wonders of nature, in which I lament at having the wrong camera lens. Barb said, "I wonder what the Indians thought when they first walked up to the edge of this canyon."

> **"I considered stopping at a supermarket and propping a frozen turkey against the gas pedal. That way my feet would stay cool and dinner would be ready by the time we got to Durango."**

"Probably, 'Wait'll the white man sees this. The eviction notice is practically in the mail.' "

On the winding road out of the park we came up behind our nemesis, a parade of big dumb lumbering motorhomes. I noted with interest that these things all have names to suggest speed and grace, like Flying Arrow, Golden Eagle and Apache Brave.

We passed through the Hopi Indian Reservation and the beautiful red pastel rock formations of the Painted Desert, which was almost as hot as the Mojave Desert but slightly dustier. The heat in the Westfield's footwell at noon was as-

tounding. I considered stopping at a supermarket and propping a frozen turkey against the gas pedal. That way my feet would stay cool and dinner would be ready by the time we got to Durango.

Just across the Colorado border we pulled into the town of Cortez right on the tail of a huge parade passing down Main Street. It was Frontier Days or Yahoo Centennial Days or something. We passed through just after the parade, and the street was still lined with people, several of whom were sober. We drove down a 3-mile gauntlet of whistling and shouting citizens. Most, I'm sure, thought we were part of the parade, a late entry. We smiled and waved at everyone, and Barb said, "I wonder if I should sit on the back deck of the car and blow kisses."

I looked at some cowboys who had just staggered out of a bar and said, "Better you than me." We crawled through town and emerged on the other end with the temperature gauge pegged, inhaling vapors of hot horse apples sizzling on an overheated Austin sump.

Climbing back into the Colorado Rockies, we made it to the old mining town of Durango by sunset and checked into the beautifully restored old Strater Hotel. We put our luggage in the room, which was done up in grand 19th Century style—hardwood wainscoting, brass bed, stone pitcher and bowl, etc—and went out for a Mexican dinner. We came back later and watched *The Thing* on our 19th Century color TV.

After an early Sunday morning start into a cool, slightly misty day we stopped in Pagosa Springs for breakfast at a Main Street cafe. As we sat drinking coffee, the cafe's clientele defected en masse to the sidewalk for a look at the Westfield. The lady at the cash register said, "I hope you folks aren't too hungry for your bacon and eggs. That's the cook out front lookin' at your car."

People driving home from church were double-parking their cars and pickups, jumping out in their Sunday best suits and cowboy boots to look at the car. We went out and joined the throng.

It was hard to know what to call the car when people asked. Westfield/Lotus Eleven was a little unwieldy, so we usually picked one or the other and said it was made in England. The engine was another problem. If you told people it had an MG engine, they'd whistle and say, "Whoo-eee! I bet that thing really flies!" If you admitted it was an MG *Midget* engine, they'd say things like, "Ha! Probably smaller than my kid's Suzuki dirt bike engine. Bet she gets good mileage, though." Caught off guard, we found ourselves describing it, variously, as an MG, MG Midget, 1275 MG Midget, Austin, Austin A-series 1275, Austin-Healey, Sprite or BMC en-

gine. For the sake of consistency, Ba and I collaborated halfway through t trip and agreed to call it an Austi Healey engine, as that seemed to pr voke the loudest murmurs of approval.

East of Del Norte we cruised steadi downhill through an ever-widening v ley and I had the feeling we were abo to be spit out of the mountains of t gold- and silver-mining West onto t prairie and cattle ranch West. The c was humming along perfectly and it w still spring in the mountains. The co tonwoods and aspens had that lacy gree look and the meadows were in bloom. was odd, here and there, to see rustic o ranch houses with gigantic parabolic T receivers in their front yards, like a Eighties update of Gene Autry and th Radio Ranch. It looked all wrong.

Near Monte Vista we hit a duststor that looked from a distance like a mo ing river of fog about 600 ft high. W drove into it to find big pieces of thing blowing across the road, mingled wit the stinging sand. Tumbleweeds went b so fast they weren't even tumbling. W passed an outdoor movie theater and explained to Barb that in order to hit th screen the projectionist would have t aim 20 ft to the left to correct for win "And on a bad night," I added, "th movie ends up in Kansas."

The river of sand died away so w took our helmets off and put our caps and sunglasses back on. It

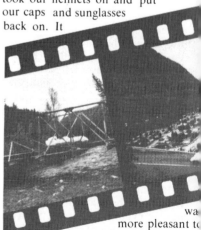

wa more pleasant t drive without helmets, so we reserved these for rain and gravel storms

On the other side of Alamosa w passed a hitchhiker who was walkin along the highway with his thumb out He was so drunk that a gust of win caused him to weave and totter into th road ahead of us. I swerved and nar wly missed hitting him. He was wearing a stars and stripes backpack, a great big felt hat with a pheasant feather in the band, a fringed leather vest, bellbottoms with decorative stitching at the cuffs, and he looked about 48 years old, sunbaked and hard as nails. One of those unfortunate ramblin' alcoholics who's inherited all the unwanted regalia of the not-so-recent past. I wondered what would happen to him. It's a very long stagger to the next town anywhere in eastern Colorado, and hard to get picked up when you're losing a drunken battle with the wind.

Highway 50 took us into Kansas, following the green willows and pretty towns along the Arkansas River. I'd been through only the corners of Kansas before, so most of my preconceived notions of the state were based on the opening moments of *The Wizard of Oz*, with its flat, sepia-toned landscape and evil tornadoes. In the early summer, at least, the Kansas we saw was a farmland of rolling green contours interspersed with small towns that radiate a wonderful quality of permanence and dignity. Our route across Kansas was one of the most beautiful parts of the trip.

The presence of evil tornadoes, however, is no myth. We stopped for lunch in Jetmore, Kansas and noticed some very dark thunderheads building up in the northwest. I asked a farmer at the cafe if tornado season was over and he said, "Nope. Just getting into full swing." As we left town a yellow biplane crop-duster swept across the highway in front of us, flying against a backdrop of dark clouds and distant pitchfork lightning. Hitchcock would have loved it.

I looked around at the sky and thought, this, by God, is real weather. Not like in California where it just kind of creeps in and sits on you. In Kansas you watch the weather at work around you like some kind of big unpredictable machine where thunderheads and shifting winds are the moving parts. Advancing clouds have a relentless quality; you can look around the broad horizon and see several storms developing at once.

Unlike the mountains, where you are often locked into a single route of travel, the plains are mapped out in a great grid of highways, so by zigzagging across the state you can play checkers with the weather and miss the worst storms. As in checkers, of course, sometimes the weather jumps you and wins. And in Kansas, sometimes the weather comes along and blows all the checkers right off the table.

At five o'clock in the afternoon we ran out of zigzag options and found ourselves in the middle of a blinding thunderstorm illuminated by tree-ripping lightning bolts. It hit us so hard and doused us so completely we didn't bother to get out any rain gear. I just put my foot in it and headed to the closest motel in Junction City. There we turned our room into a vast drying rack for clothes, sleeping bags and tools from my water-filled toolbox. In heavy rain, it turned out, the rear tires churned great quantities of rainwater up into the rear bodyshell, so that it ran down our seatbacks like a waterfall and filled the car. I added, "5. Needs inner fender wells" to my improvement list.

The morning after the storm was cool and clean with a sky the color of morning glories. Willard Scott, on our motel TV, said there'd been 25 tornadoes in the plains states the previous day, with more on the way. As we headed across the Missouri River, however, there wasn't a cloud in the sky.

Northern Missouri was our first exposure on the trip to real sports car country. Narrow winding roads, hills and steep valleys, 1-lane bridges and trees forming a tunnel of shade over the road. It was also the most dangerous section, because farmers with tractors and manure spreaders were not ready for us. Nothing in their lives had prepared them for the specter of a small white bomb of a car with blue racing stripes to come drifting around a blind corner or bearing down at high closing speeds. We weren't driving too fast for the Westfield; just too fast for pickup trucks and hay wagons with a relaxed indifference to the centerline. After a few close calls we decided to slow down and live to enjoy the scenery. The Westfield was in its element, running cool and handling the road flat and quick as a go kart.

On Highway 6 I saw a sign for a place called Novinger and turned off the road, down the town's main street. My parents had lived in Novinger for a couple of years before I was born. My dad had bought a small weekly paper there with his Navy savings at the end of World War II. I'd heard so many stories of this little coal mining town that it existed larger in my imagination than the places I'd seen myself.

It turned out the coal mines are closed now and the only people living in town are those who work elsewhere. We drove down a ghost town main street of dusty, boarded-up storefronts, peeling paint and faded signs. It looked like the town where Bonnie and Clyde tried to rob the bank and found it closed. A heavy-set young man with, believe it or not, a Mohawk under his seed cap and a T-shirt listing Ten Excuses for Not Having Sex gave us a tour of main street and pointed out where all the stores had been. He couldn't remember a newspaper office ever having been in the town. We stopped for a drink in the one open bar, and no one in there remembered either. Novinger was like a lot of country towns we passed through. It was a little too small to support commerce, so people just gave up on the idea of having a town and quietly moved away. What remained was a house collection, with bar.

Driving into southern Iowa at sunset we passed a slow semi on a long hill. I casually looked up at the driver as we passed and was greeted with a blinding flash of light. As my vision recovered I realized the driver was grinning and waving a small Instamatic flash camera at us. He gave us the thumbs up sign and backed off so we could pass. As we pulled ahead another flash was fired at our backs. We drove into Ottumwa and found a motel with blue dots all over it.

In the morning I checked the oil. It was down a half-quart, after 2000 miles of driving in hot weather. Our best mileage up to that point had been 52 mpg (Kansas tailwind) and our worst had been 43 mpg (Colorado mountain headwind). None of the nuts and bolts I checked had loosened up, and after five days the car was running cool and strong. Valve clearances hadn't changed since Williams, Arizona, and the plugs were a nice tan color. The points looked good, timing was spot-on and the oil pressure was still at 55 psi hot. I'd left California fearing the Westfield might be a slightly fragile, fussy sort of car on a long trip. Now, a few hours from the Wisconsin border, I'd begun to think of it as a remarkably tough, durable machine. Most of my constant listening, checking and adjusting had been wasted effort. As we slid into the car on the sixth and last day, it was like putting on a comfortable pair of shoes. The sound of the engine starting on that clear summer morning was pure music.

We crossed the Mississippi near Dubuque at mid-morning and headed into the green Ozark-like hill country that is southwestern Wisconsin. My growing up in that area has nothing to do with a personal opinion that this is some of the most beautiful country on earth. It was shaded valleys, rivers, red barns and villages all the way into Madison. From there we drove the back roads to Elkhart Lake, passing through pretty towns like

Greenbush and Glenbuelah, and all of a sudden we found ourselves braking and downshifting into the city limits of Elkhart Lake, something I'd first done 18 years ago in my green TR-3. We motored down the shaded village streets, past the lawns and white porches of Siebkin's Hotel and headed south toward the gates of Road America.

On the seventh day the Westfield rested beside the track. Chris Beebe and his Lotus Super Seven did all the work, winning the D-Production race in a hard-fought battle. About 30 of us who like to celebrate such things retired to our traditional campground on the

shores of Lake Michigan, pitched tents, ate grilled bratwurst and drank Bohemian Club, the official beer of southwestern Wisconsin Lotus Seven owners and their ilk. In the morning all evidence suggested we'd had a good time.

When the race weekend was over, Barb had to fly back to California and her job, so Chris Beebe agreed to drive the Westfield back across the country with me. Before leaving, we worked on the car for three days at Chris's shop in Madison. We made a new heat shield for the carbs, ducted air to them, moved the radiator forward in the nose and added shrouding, bolted a piece of asbestos to

the driver's footwell, installed some inner fenders over the rear wheels and jacked the front ride height up with taller springs. (Several of these improvements have now been made by the Westfield people.)

The return trip was pleasurable and uneventful, and we made the 2200 miles without having to lay a hand or wrench on the car. Our small mechanical improvements worked wonders, eliminating the Westfield's few irritants. The engine ran cool, the rear SU no longer overheated, the footwell was incapable of roasting a turkey and the oil pan

Continued below

Continued from above

didn't bottom on gum wrappers and lost coins. Chris had a great time and said he hadn't seen people so intrigued with a car since he and his father took a trip in an XK-120 Jaguar in the early Fifties.

The best piece of road on our return route was a stretch in the Colorado Rockies, 110 miles of mountain switchbacks and hairpins between Montrose and Durango on Highway 550. When we pulled up in front of our hotel in Durango, Chris said, "I've been trying to think what car might have been more fun to drive on a road like that, and I can't think of any."

I couldn't either. The Westfield had some of the oddball charm of the TC Chris and I had driven to Road Atlanta the year before (R&T, April 1983), but it was faster and the steering worked. The racing car heritage, too, had an appeal. Looking over that low windscreen at the sleek front bodyshell, sitting in a stark aluminum interior and listening to the raspy exhaust note as you downshifted into a corner, you needed a firm grasp on reality to remind yourself that this was *not* the last Mille Miglia. A helmeted passenger sitting next to you with a map on his knees did little to dispel the image.

The Westfield was just outlandish

"On the seventh day the Westfield rested beside the track."

enough that you didn't mind any discomfort it dished out. Other sports cars had become more civilized by small degrees until they were so much like sedans that people could no longer remember why they wanted a sports car in the first place. No one, if he were honestly building a car for sport, would load it down with carpets and courtesy lights and 40 pounds of window-winding mechanism. Maybe, I suggested to Chris, sports cars really are supposed to have nothing added that doesn't make them go fast, and every so many years

someone has to rediscover that. Like the people in England who make the Westfield.

When we got back to L.A. late on a Friday afternoon and finally pulled into our driveway, we were neither relieved nor happy to have the trip over. Tony Hogg, rest his soul, had been wrong when he said the Westfield trip would be just bloody awful. He was only kidding, of course. He had spoken mainly for effect, with a glint of good humor in his eyes. Tony himself had been a true enthusiast of everything automotive that is pure fun. He knew better than anybody that driving an open sports car across America's back roads in summer is further from being awful than nearly anything. ⊚

MOST TEST CARS arrive at our palatial Newport Beach offices in one of two ways. Either the manufacturer drops the car off on our doorstep or we go to the manufacturer's distribution center and pick up the car ourselves.

Very seldom does a test car arrive completely disassembled on three separate forklift pallets in the back of an airfreight delivery

vince the Good Editor, the Beneficent Publisher and the Kindly Business Manager that what this magazine needed was a really good kit car, to be assembled as a magazine project. It worked.

The very words "kit car," of course, may be enough to send shivers of dread up and down the spine of the average hard-core car enthusiast. They conjure up immediate images of a disrepu-

CRATE EXPECTATIONS

Building Westfield's Lotus 11 replica

BY PETER EGAN
PHOTOS BY THE AUTHOR

PHOTO BY BARBARA EGAN

Road & Track's sophisticated new forklift at work on arrival day.

Driver's footwell came filled with trim pieces and ancillary items.

truck. Equally seldom does half the staff stay well past quitting time on a Friday night to greet a car and lift its various and weighty components from the back of that same truck (yes, we have no forklift). In fact, these things have never happened at all, until recently.

Blame for this sudden madness can be placed squarely on the shoulders of three people. First, the late and brilliant Colin Chapman, who in the mid-Fifties designed a lovely aerodynamic sports racing car called the Lotus 11. The second guilty party is an Englishman named Chris Smith, whose Westmidlands firm has long specialized in the restoration of vintage Lotus racing cars. It dawned on Smith that it would be just about as easy to build an entirely new Lotus 11 from the ground up as to restore some of the bent and rusted iron that came through the front door of his shop, so he set up production facilities to produce a Lotus 11 replica known as a Westfield. The third culpable individual is R&T's own English correspondent Doug Nye (curse these clever Englishmen) who wrote a glowing report about the Westfield sports car for our June 1983 issue and gave all of us a vicious dose of the dreaded English Sports Car Lust, a disease many believed had been eradicated by modern technology.

In other words, after reading Doug's article on the Westfield car, we all wanted one. Being eyeball deep in canceled checks from my Lotus 7 restoration project (talk about bent and rusted iron), I decided the cheapest solution to this craving was to con-

table parody of some nice old British roadster tarted up with too much upholstery, fake louvers, the wrong steering wheel, phony exhaust plumbing, a picnic basket where the engine should be and a tell-tale set of Volkswagen mufflers whistling from beneath the rear deck. There are a few nicely done kit cars around, but most tend to misinterpret the spirit of the original classic, or cost more than the real thing. Or both.

In reading Doug Nye's report, however, a picture emerged of a nicely crafted, eminently affordable kit that stuck very close to the original design and avoided many of the usual replicar pitfalls. With its tubular steel space frame and riveted aluminum chassis panels, the Westfield is very close in design to the original 11. It differs only in the use of fiberglass for the upper body panels (molds lifted from the original Lotus aluminum), heavier-duty tubing and a few extra braces in the frame to handle the rigors of street and pothole, and a slightly lengthened cockpit section to accommodate tall drivers and the Sprite driveshaft.

Though various engines can be adapted, the basic Westfield kit is designed around that widely available junkyard phenomenon, the drivetrain from a rusted-out Sprite or Midget, preferably the 1275-cc version. A live rear axle, also Sprite, is located by a Panhard rod and four trailing links. The front suspension uses Spridget spindles, steering rack, hubs and disc brakes, but the suspension pieces are fabricated Lotus-like from steel tube. As on the original 11 (and 7) the front anti-roll bar doubles as

One Westfield sports car, laid out and ready for assembly.

the front half of the upper A-arm. Coil springs and adjustable Spax tubular shocks are used front and rear.

Like most early Lotus cars, the 11 was always available in kit form with a variety of proprietary engines, so a modern-day kit seems like an extension of—rather than an affront to—tradition. The Le Mans model was powered by the 1100 Climax engine and had De Dion rear suspension, but the more affordable Club model used a live rear axle, and the still cheaper Sports version used a live rear axle with the inexpensive side-valve Ford 100E engine. The original 11 also came either with or without the aerodynamic hump behind the driver, and Chris Smith offers the Westfield in both body styles.

The U.S. distributor for Westfield is Rev-Pro Engineering Inc in Sarasota, Florida, another restoration and racing fabrication shop, owned by a very helpful gentleman named Willard Howe. Rev-Pro (6223 S. McIntosh Rd, Sarasota, Fla. 33583; 813 922-7371) sells the cars in full or partial kits, or as completely assembled vehicles, depending upon the customers wishes and checking account. All mechanical components are reconditioned. Prices range from $4600 for a basic chassis kit sans running gear and other Sprite pieces, up to $10,500 for a fully assembled car. At the time we decided to go ahead with the project, all Rev-Pro's cars were spoken for, so we called Chris Smith in England and found that he had an unsold left-hand-drive car with blue upholstery nearly completed on the assembly line.

By the time we ordered the car I had already tracked down a really clapped-out 1971 MG Midget ($400, with crash damage and no discernible oil pressure) that I planned to rebuild for our drivetrain. Chris Smith informed us, however, that his price for a freshly rebuilt engine, transmission and rear axle was only $1100. That was less than it would have cost to buy and rebuild the innards of the $400 Midget, so we ordered the complete kit for a total cost of $7100. We requested the car be airfreighted, as I hoped to have it built in time to drive to the June Sprints at Elkhart Lake, Wisconsin. If you think it's expensive to airmail a Christmas card to your aunt in London, try sending a car. The postage figure is best left unmentioned. Take our advice and let Willard Howe get you one by boat. Have patience and it shall make you rich.

Sprite rear axle uses existing brackets for shock and trailing link installation. Only modification is stud for Panhard rod.

Three weeks after placing our order, the car was at the door. The Westfield actually arrived as three separate shipments; one for the chassis parts, one for the 1965 MG Midget engine and transmission, and the last containing the rear axle assembly. The largest pallet, of course, carried the chassis and primer-gray fiberglass bodyshell. This arrived covered with plastic wrap, criss-crossed with yellow tape reading "FRAGILE" and filled to the gunwales with random Westfield parts.

Unloading the pieces from the chassis tub was about as close to Christmas morning as anything I've experienced since the earlier part of my arrested childhood. The driver's and passenger's footwells were filled with items such as red Maserati air horns, carefully wrapped plexiglass windshield pieces, throttle and choke cables, exhaust header, red leather-covered steering wheel, a BSA motorcycle muffler, hood latches, bags of nuts and bolts and other pieces whose mysterious functions became clear only as assembly progressed.

I was surprised at the completeness and finish of the chassis when we got the miscellaneous parts cleared away. The basic chassis arrived fully painted and upholstered, with the dash (instruments mounted) and all the aluminum panels riveted in place. The front suspension was bolted loosely together, and the pedal assembly, steering rack and column were fitted. The engine, transmission, rear axle and brakes were all clean, freshly painted and—from all appearances—thoroughly reconditioned.

Not one printed word of instructional material was included with the kit, but unless you're the sort of person who accidentally bolts exhaust headers to the rear axle, the relative position of parts is pretty self-evident. The Westfield is basic automobile at its best.

PHOTO BY STEVE KIMBALL

I decided the most logical progression was to get the car on its wheels with the brakes hooked up, install the wiring harness, slide in the engine, transmission and cooling system and get the car driveable before dealing with the bodywork and cosmetics.

Assembly of the rear suspension is reasonably easy. You center the rear axle in the frame and then shim the trailing links with washers so they move freely with the axle in its proper location. With the axle shimmed the Panhard rod drops right into place and the handbrake assembly can be hooked up. Front suspension, as mentioned, was already bolted loosely together, but there was more involved than just tightening down the bolts. Because of differences in manufacturing tolerances and weld thicknesses, some of the suspension bolts were too short for correct bedding of the locking rings in the Ny-lok nuts, so I did a bit of juggling with washers and bolts, filing the surfaces and substituting with bolts from my own massive and motley collection of project leftovers (the Westfield now has a grade-8 Lola trailing link bolt holding its left lower A-arm in place).

In the never-assume-anything school of auto mechanics, I took out all the factory-assembled nuts and bolts, to inspect and reassemble them with thread locking compound. If you aren't in a hurry to get the car ready for a trip to Elkhart Lake, it is probably also a good idea to drill and safety wire some of the more crucial suspension bolts. I didn't, however, and nothing loosened up after the car was assembled and driven.

Assembling the brake system was time-consuming because the flare fittings were new and unbedded and they all leaked after the initial tightening. Some sealed after considerable loosening and retightening, and others needed burrs touched up with light emery paper. The front calipers have double pistons and were very difficult to bleed because air wants to stay trapped in them. The calipers had to be removed from the spindles and rotated at various odd angles during bleeding (with a steel spacer between the pads) to get the air out. Even now, the pedal is not quite as hard as it should be, and I suspect there's a renegade bubble in there somewhere.

I greased the hub splines and mounted the brand-new 14-in. MGB wire wheels and Ceat Veltro radials, tightened the knockoffs down with a lead hammer and lowered the car to the ground. A roller.

At the time our kit was shipped, Westfield had not completed production of its wiring looms, and part of the deal was you had to fabricate all your own wiring. Chris Smith knew we were short on time, so he kindly tossed in a stock wiring loom out of a late model MG Midget. I spent two nights going blind on wiring diagrams and trying to track down wires that led to nonexistent 4-way flashers and seatbelt buzzers before throwing the MG wiring loom at a nearby stack of empty beer cans and cigarette butts. I called two friends, John Jaeger and Roger Salter, both of whom have restored many old British cars and profess actually to *like* making up wiring looms. (Too much Guinness. Or not enough.) At any rate, they're very good at it and fabricated a beautiful harness while I used the R&T floor hoist and a chain to drop in the engine and transmission. The drivetrain went in effortlessly, once I put the motor mounts in right-side up.

I adjusted the valves, installed the distributor, set the static timing, added oil, hooked up the battery, cranked the engine for oil pressure (70 psi cold), installed the plugs and twisted the key. The engine fired immediately, settling down to a smooth, mellow idle through the BSA bike muffler. I took the car out for a midnight drive and the throttle immediately stuck wide open (minor adjustment). Otherwise the car ran, felt and sounded beautiful. Without the weight of the bodywork it felt like a go kart and accelerated like hell, almost leaping down the road under bursts of acceleration.

In its standard position, the clutch pedal had too little travel to fully disengage the clutch, so I got my welder and made an adjustable pushrod for the clutch master cylinder. Those with shorter legs could accomplish the same thing by moving the entire pedal cluster closer to the driver by drilling a few holes in the pedal support bracket.

We sent the bodywork out to a paint shop while the fine mechanical details were being looked after. I've painted at least a dozen cars myself and now have lungs I imagine to be about two thirds filled with Bondo, primer dust and paint spray, so I prefer to farm it out these days. We chose a soft white paint with a medium blue racing stripe down the middle of the car. The color was chosen partly for tradition, because R&T had a Lotus 11 in American racing colors on the cover of the March 1957 issue when we first tested the car. The other reason was because the fiberglass on the Westfield is a bit bumpy and cobby and reflects fewer of its flaws in a lighter color. A dark green, for instance, would demand a lot of sanding and preparation.

The final project was to install the windshield, doors, side windows, headlights and covers. Most of these pieces come undrilled, so careful positioning is required to make everything fit before the holes are drilled. The kit provides polished brass nuts and bolts for windshield attachment, which add a nice finishing touch. The headlight covers needed some moderate shaping on a belt sander to fit their recessed openings, and the quartz-halogen headlamps provided were too long for the headlight buckets, so we installed a pair of smaller 6-in. motorcycle headlamps under the covers. The lights and all the wiring worked, the horn honked, and the car was done.

As kit cars go, I would say the Westfield has just about the right balance. It is not so thoroughly pre-assembled that anyone can just slap it together with acceptable results, nor is it so difficult that someone with reasonable mechanical skills and a good set of tools—and plenty of spare bolts and washers—can't produce a pleasing, well finished car. That, of course, is the appeal of a kit car. No two will come out quite the same, and the car becomes an expression of the builder's patience, craftsmanship and vision. The Westfield is a sound, well designed car that holds no unpleasant surprises for its constructor, but demands just enough careful massaging and detailing to make the construction process interesting and personal. When you're done, you feel as though you own this car in a way no one else can.

Working nights and weekends, I was able to finish the project in just over three weeks with a little help from my friends. That, of course, does not include the sanding and painting of the fiberglass bodyshell, which was done by a paint shop to the tune of $600. So with the cost of our kit, the paint work, and random nuts, bolts, fittings, fluids, wire and tape, the total cost of the car (excluding airfreight) was well under $8000. Expensive, but not bad these days for a new sports car, and cheaper than many other kits and replicas on the market. A buyer could save money, of course, by purchasing the basic chassis kit and then hunting down his own drivetrain, instruments and ancillary items.

How does the car work? Just fine, thank you. It's quick, light, fun to drive, and handles so well you can't believe the suspension was designed nearly 30 years ago. (But then the man who designed it was good at this sort of thing.) Suffice it to say that when the sun is just rising on a Sunday morning and you just happen to own a cap and a pair of goggles, the Westfield is not a bad thing to have lurking in your garage.

DINO
CAR OF THE NORTH

In winter, a crooked line from Wisconsin to California is the coldest distance between two points

BY PETER EGAN

PHOTOS BY THE AUTHOR

"**P**ETE, HAVE YOU ever seen the Badlands in the dead of winter?" It was a bad long-distance connection, one of those crossed-up fiber-optic wonders where you can hear a woman from Tulsa discussing her kidney operation in the background. Still, I knew it was my old friend Chris Beebe on the phone. I used to work with Chris at his foreign car repair shop in Madison, Wisconsin, and he's the only person I know who opens conversations with a non sequitur when no premise has been suggested.

"Why no," I said obligingly, "I've never seen the Badlands in the dead of winter. Why do you ask?"

"A friend of mine named Carl Maguire has a car collection here and he's moving to San Diego. He asked if I'd like to drive his 1972 Dino 246 GT to California, and of course I said yes." (A pause here for the sinking in of thought.) "I couldn't help thinking that it seems like a long drive for one person. It might be a nice opportunity for a California journalist to drive a sports car in real weather, like the rest of us have to . . ."

"How's the weather out there?" I asked.

"Terrible. They're saying it's the coldest midwest winter in more than 100 years, and one of the worst for snow. Hundreds of people have died. It's 10 below zero right now, but it's supposed to warm up to zero tomorrow and snow again. On the positive side, if you left right away you could be here in time for the Friday night fish fry at the Old Stamm House."

A few days later I was on a Frontier Airlines DC-9, watching the sunny Mojave Desert turn into the Rockies and the snow-covered steppes of Nebraska. I spent most of the flight paging through Doug Nye's excellent book *Dino, the Little Ferrari*.

Though I'd never driven a Dino 246 before, nor even sat in one, I imagined it to be my favorite Ferrari (okay, along with the 250LM). There were bigger, sexier and more exotic Ferraris, Ferraris with fire-breathing 12-cylinder engines, and some with much greater historic and collector value, but few that attracted me personally as much as the Dino and

DINO

its high-winding 2.4-liter V-6 engine.

Perhaps a good part of the appeal was accessibility. It was the only Ferrari I desired that ever appeared on the used car market at prices a man could nearly afford if he sold every last thing he owned, except for his old British sports cars. It was that almost-within-reach quality that made the Dino especially attractive.

Furthermore, I loved the shape of that Pininfarina body. With its windshield curved into the hood P3 style, rounded fender lines, wraparound rear-window glass and air scoops behind the doors, it sat on its wheels with a squat aggressiveness that made it appear more alert and tossable than the larger Ferraris. Chunky and sleek at the same time, it looked sculpted rather than stamped, predating the odd, arbitrary notion that beautiful cars are designed by folding and creasing flat sheets of paper.

I knew it was cold in Madison by the amount of steam coming off the roof-

tops. Chris and another old friend, George Allez, picked me up at the airport. Snow flurries were blowing around as we drove out to the Old Stamm House for its famous Friday night fish fry, then back to Foreign Car Specialists for a look at the Dino.

The car was silver metallic with a tan interior and appeared to be in very good shape. "How does it run?" I asked Chris.

"Pretty well, considering it's been in storage for a long time. Carl has barely had a chance to drive the car since he bought it. The heater isn't working, so we'll have to check that out. Also, the steering feels a little twitchy, so it'll probably need an alignment. The oil pressure reads zero. Broken gauge, I hope."

We spent Saturday changing oil, checking tire pressures, belts, etc. The heater, it turned out, was working, but one of the fresh air vents was stuck open, diluting the heat with subzero blast from the great outdoors. We closed the vent and duct-taped cracks and seams in the heater hoses and footwells. Carl Maguire stopped by and I finally got to meet him. Carl is a surgeon and his other cars are a Lotus Europa and a Lamborghini Miura S. These two, in need of various repairs, were going to California by trailer. Carl seemed like a very nice fellow, but then I always enjoy meeting people who have as many automotive screws loose as I do.

We arose on a gray Sunday morning at Chris's farmhouse and sat down at the breakfast table to consider our route. Chris's road atlas, which he refers to as "the Rand-O'Malley," is apparently some kind of family heirloom, carried around these many years in the same shoebox with his childhood baseball cards. It was already badly out of date when the Milwaukee Braves won the pennant. Interstates and many other modern highways are not depicted because the settlements they serve hadn't yet been founded at the time of publication. That was fine with us, of course, because we both subscribe to the Persig theory that the most direct route between any two cities is always the least interesting, and Interstates are out of the question.

Looking at the U.S. map, we decided to take the rugged northern route for the sheer adventure of battling the worst winter in over 100 years, rather than copping out and heading straight south. Our master plan was to drive through Iowa "because it's in the way," South Dakota to see the Badlands in winter, and Wyoming, Utah, Nevada and California for the mountain roads.

We gunned out of the driveway at 10:00 a.m., having packed the car with luggage, a wool car blanket, a few tools and a red plastic child's snow shovel with a single staple holding the handle

on. I had prepared for winter emergencies by stuffing my suitcase with more wool sweaters and socks than I could possibly wear at any one time. These were packed in the small trunk behind the Dino's engine compartment. We took a series of winding county roads to Prairie du Chien and the Iowa border.

By 10:30 the engine was fully warm and it became apparent this was going to be a very cold trip. I put the heater on full hot and the fan on the highest of its three settings. Taking off one glove, I put a hand over the passenger-side heater duct.

Chris looked at me expectantly. "How is it?"

"It feels like a hamster blowing through a straw."

With the system on full defrost, the windshield stayed clear in two fan-shape arcs, but the footwell was cold as a grave, though draftier. By noon we had labeled the heater fan positions as follows: COLD?/YES!/NOT SO HOT/OFF.

Neither of us complained. We had somehow expected it. Ferraris, we reasoned, were never intended for midwest roads in the dead of winter. They were built to sweep through sun-drenched Sicilian villages in the Targa Florio, places where old men sat at tables beneath the olive trees, drinking grappa and saying things like, "It is good to go very fast," or, "These young ones, what do they know of death?" And here we were, cruising through a frozen Iowa farm town where the sign on the Farmers & Merchants Bank said "+3°" and "Have a Nice Day."

Still, a working heater would have been nice. My dad's Pinto had one.

At about 2:00 p.m. my pie-and-hot-coffee light came on, so we began to look for a small town cafe. The restaurants in one town after another were closed, however, and we realized that Sunday afternoon is a bad time to get hungry in Iowa. Decent folks are home watching football, not prowling from one state to the next in borrowed Ferraris. We finally spotted a cafe with the lights on, parked and walked up to find the door locked, with a high school girl vacuuming the floor inside. "Closed," she shouted over the roar of the vacuum cleaner. We pulled back onto Highway 9 and pressed on.

The Ferrari's handling up to this point could best be described as self-slaloming. We hadn't got around to the alignment, thinking we could get that done on the road if necessary. The steering was acceptable, if a little twitchy, on dry roads. But when we hit ice patches on the highway, it demonstrated the effects of extreme toe-out. If the right front tire hit ice, the grip of the left took over and jerked the car toward the oncoming lane. Ice under the left tire sent

DINO

us toward the shoulder. The left front wheel was also out of balance, lending a light jackhammer effect to the steering wheel. Chris looked at the darting, shuddering steering wheel and said, "We've got to get this fixed."

In central Iowa it began to snow heavily out of a dark afternoon sky. We turned on the radio for a weather report. I hit the scan button and the scanner needle went back and forth across the dial three times without stopping for

anything. "No radio stations in Iowa?"

"Maybe it's a discretionary scan," Chris suggested, "and there's nothing good on."

Half an hour later, we didn't need any weather reports. There was a blizzard on the plains and we were in it. The snowplow brigade was out in force, trying to keep up with a dry, heavy snow blowing out of the south. The Ferrari was anteating all over the road, as Henry Manney would say, Chris working the wheel to avoid oncoming plows that appeared out of the dark in a swirl of light and snow.

As we pushed blindly on, our headlights caught a city limits sign for Clear Lake, Iowa.

"Clear Lake," I said. "This is the town where Buddy Holly played his last gig. His tour bus broke down, so Holly and a couple of others chartered a plane to fly to their next club date in Minnesota. The plane crashed in a field just north of here. The Big Bopper and Richie Valens were killed, too," I added. "It was a night just like this; snow, poor visibility . . . It would have been about 25 years ago this week. I remember, because he was killed just before my birthday . . ."

I was rambling on in this slightly macabre vein when a farmer who apparently didn't see us coming suddenly caught traction while spinning in his driveway and launched his Ford LTD onto the highway in front of us, blocking both lanes. My only reaction was to mutter "Oh no" and brace myself for the impact. There was no room to stop on the icy highway. He was going 5 mph and we were going 50.

Chris downshifted a gear, got back on the power and blasted through a snowdrift on the shoulder, sliding half sideways around the LTD and back onto the road without touching the car or the 6-ft wall of plowed snow near my elbow. I had a glimpse of the other driver, looking back toward his farm, and I don't think he ever saw us. The blowing snow swallowed us up and his headlights disappeared in the mirror.

"That was very smooth," I said to Chris a few minutes later. "Dinner is on me."

By the time we reached Estherville, Iowa, the weather was ridiculous and we were relying on a 4-wheel-drive pickup truck to blaze a trail through the snow ahead of us. We picked out a dim neon MOTEL sign and wafted across the parking lot, splashing and spinning through drifts. When we pulled up next to the office, I couldn't tell if Chris was parking or just stuck. The motel manager said it was good we had stopped because all roads out of town were now closed. "We got a bunch of semis off the road or snowed in," he said.

We awoke to a clear, windy morning, storm gone, roads plowed and car only partly buried. A short dig and we were on our way.

After crossing into South Dakota we stopped at a gas station to put some heat-preserving cardboard in front of the radiator. Three mechanics came across the highway from a nearby farm implement dealership.

"A Ferrari!" the tallest mechanic exclaimed. "You don't see these except on TV. *Magnum.* Mind if we look under the hood?"

I raised the engine lid, and while Chris and I worked up front on the radiator, the tall guy waved his arm around the engine compartment, pointing out important features to his two silent colleagues. "Look here," he said, "three carbs, gas turbine . . . fuel injection . . . This baby's got everything."

When they were gone, Chris and I went back and looked at the engine to see if we had missed something.

We crossed the Missouri River, which I always think of as the dividing line between Midwest and True West, driving through towns with terse names like Tripp and Platte, places where the city fathers were either too hot or too cold to bother with a second syllable. Cruising into Winner, South Dakota we decided an alignment was long overdue. After hitting two shops that refused even to look at the Ferrari, the friendly service manager of a big Ford garage agreed to give it a try.

"Hey, Henry!" he shouted into the back of the shop. "You want to align a Ferrari?"

Henry, a middle-age man with the remains of an unlit cigar clenched squarely between his teeth, poked his head out of the alignment pit and said, "What the hell's *that*?"

Laughter of the har-har variety came from beneath the open hood of every pickup truck in the shop.

"Sure, bring it in," he said. "What the hell."

While Henry performed the alignment, the service manager amused himself and others by cracking a genuine leather bullwhip in all directions around the garage. We had stumbled on one of those colorful character collections nurtured by certain, select garages. Fifteen minutes later the alignment was done, only $8.98, coffee and bullwhip demonstration included.

The Ferrari was a dream to drive. It tracked straight and steered through corners with new, calm precision. Even better, the weather had cleared and central South Dakota was in the grip of an unseasonably warm spell. We cruised out of Winner with the windows rolled down, elbows on the doors like people with spring fever. A few miles down the

DINO

road, Chris suddenly turned to me and said, "My feet are warm," in the same tone of voice soldiers use when they say the shelling has stopped.

Wanting to see the Badlands before sundown, we began to push the pace. Above 170 km/h, we discovered, the alternator belt began to slip and smell like burning rubber. Not wishing to remove the right rear wheel and inner fenderwell to get at the alternator, we slowed down to 165 km/h, or about 102 mph.

The Badlands in winter were not the snow-blown wasteland we'd expected. We arrived in this strange landscape almost at dark, with the peaks and mesas outlined against a purple sunset. There was no snow, and a dry, balmy wind rustled through the canyons. We slowed to a crawl on the empty park road and drove the miles of valley road with our headlights off, the Ferrari growling along in 2nd gear like a wary animal. Slow driving made the Dino nervous and edgy, causing it to grumble and whine and make small backfiring noises.

After spending the night in the little town of Wall, we experienced the novelty of having breakfast as the only tourists at the famous Wall Drug. In the off-season, this huge restaurant/gift shop opens only one small corner for seating, right near the fireplace. We ate amid tables of local merchants, ranchers and retired cowboys bristling with cowboy hats and string ties.

We filled up at a gas station in Wall and noticed a pattern that was repeated throughout the trip. The station attendant came out and filled the car silently, regarding us and the Dino out of the corner of his eye, with a sort of aloof curiosity. "Nice car," he said at last, taking my credit card.

"Yeah," I said. "I wish it were mine. We're delivering the car to California."

On hearing that, the man opened up, began asking questions about the Dino and asked if he could see the engine. How fast did it go? What was it like to drive? Where were we headed tonight? It happened over and over again, at gas stations, motels and cafes; people became a lot friendlier when they found out the car wasn't ours. There was an intimidation factor built into the Ferrari, something Chris and I hadn't experienced in previous cross-country trips we'd taken via MG TC and Westfield. People admired the Dino but kept their distance until we went out of our way to be friendly.

Near Rapid City we hit our first unavoidable stretch of Interstate. Where the Ferrari had dominated the smaller 2-lane roads, it felt low and dwarfed by all the pavement on the I-road. In Rapid City we treated the car and ourselves to a wheel-balance job at a place called Uncle Milt's Alignment. It was satisfying to watch a few lead weights turn our juddering left front tire into a smooth, humming gyro.

Not a single tourist trap was open on the highway into the Black Hills; Wax Museum of the Stars, Parade of Presidents Museum, Life of Christ Holy Shrine Wax Museum and Nature Mystery Area were all dormant. After 68 miles of signs warning us to get ready for a great time at Reptile Gardens, the place had a giant CLOSED sign on the door. Chris and I groaned in unison, imitating a pair of disappointed 9-year-olds. We got out and had an impromptu picnic on a sunny hillside near Mt Rushmore, opening a basket Chris' friend Gail had sent along. The off-season emptiness around this tourist landmark was almost eerie, as though we were the only survivors of some general catastrophe and hadn't yet got the news.

Coming down from the hills we rolled into Wyoming cattle country, and evening found us shopping for a motel in downtown Laramie. We found ourselves assigning construction dates to each motel. You can tell how old a motel is and how recently it's been updated by what the signs promise: ice water, telephone in room, radio, heated rooms, steam heat, TV, satellite cable TV, air conditioning, pool, X-rated movies in room, free coffee, etc. It was a simpler era, I reflected, when you could lure Americans into a motel with ice water rather than X-rated movies. People made their own fun in those days.

We picked a place called the Buckaroo Motel, circa 1955, TV and coffee in room. There were plywood silhouettes of cowboys on horseback on the door of each cabin. Wyoming is a state where you see an end-of-the-trail Indian, lance dipped, or a bucking bronco on about 80 percent of all man-made surfaces.

You have to get stuck once on every winter trip, so I turned off on a private ranch road near Walden, Colorado to take some pictures and dropped a wheel into the ditch while turning around. We dug until the staple fell out of the red plastic child's shovel, then pushed and spun our wheels for effect. It was hopeless, so we walked about a mile to the ranch. We were greeted by dogs, cattle, cats and then the woman who owned the ranch, in ascending order of friendliness. She was dressed in a barn jacket and tall rubber boots, and had the good-natured, squinting smile of people who spend a lot of time in the out-of-doors. "Looks like you got a little trouble," she said cheerfully. "I'll get the hired man to pull you out with the Cat, as soon as he's done feeding the cattle."

The hired man, a fellow of few (no) words, gave us a Caterpillar tractor ride back to the car at about one quarter walking speed, with a dog in tow. We hooked up a chain and it took the Cat

DINO

about two seconds to wrench the car out of the ditch. The man worked with the unblinking efficiency of a guy who pulls 20 or 30 Ferraris out of the ditch before breakfast every morning. We paid him for his trouble and he smiled, tipped his hat and roared away.

"Triumphs, MGs, Healeys and now Ferrari," I said to Chris. "In the end, I have always relied on the kindness of strangers. Someday I'll have a vehicle that can pull someone else out of a ditch."

We descended into Utah and the frigid Green River Valley, cruising into Vernal, Utah, "The Dinosaur Capital of the World." The main street had a Dinosaur Motel, Dino's Dinah Club, the Dinosaur Inn, Dinah Bowling and a laundromat with a mural depicting happy dinosaurs doing their laundry. Chris said, "I wonder if they get many Dinos around here."

"Extinct," I said. "Too cold for them."

On the other side of Vernal the cold took on a new intensity, the kind that creeps around the doors and sinks into your bones. The sun slipped behind some clouds, and the Ferrari, which depends heavily on greenhouse effect for its heating, felt like an animal losing body heat too fast for survival. I took off a glove and held one hand over the defrost vent. "Remind me to beat the hamster when we get to Provo," I said.

Chris took off his boots and put three pairs of wool socks over the two he was wearing. When that didn't help, he pulled a pair of wool gloves over his feet, which made him look like some kind of Dr Seuss hero. At that moment a Utah State Patrol car pulled us over for speeding. Seventy-one mph.

The cop listened to our explanation that we were trying to drive fast to get to a warmer state, looked at the gloves on Chris' feet and then wrote the ticket. He let us sit in his car to warm up and advised we pour pepper in our socks at the next restaurant. "It sounds crazy," he said, "but it really makes your feet warmer."

We tried it at dinner that night in Heber City: one foot at a time for comparison purposes. Neither of us noticed a difference, but then our feet were already numb. The waitress watched from a safe distance as we poured pepper into our socks, and I could tell from her expression she thought we were a couple of really neat guys.

The next day found us driving into the Valley of Dry Ice; the west ridge of the Uinta Mountains had trapped a layer of moisture, and sub-zero temperatures had turned it into a strange theatrical fog. We meandered endlessly in the mist in a cluttered suburban area around Spanish Fork, trying to find Highway 6 south of Salt Lake City. We finally found our highway and drove out of the cold and fog near the Nevada border.

West of the Snake Range, Nevada was all sun and warmth, with long, sweeping ascents and climbs on straight, empty

DINO

roads. As we accelerated out of the little mountain town of Ely, Chris ran the engine up to redline in each gear, leaving a wonderful high-pitched snarl in our wake. He smiled and shook his head. "Whenever we drive out of a town, this car makes me think we're being timed."

I knew the feeling. There was something about the Dino that added a sense of urgency to your driving, a feeling that some unseen race official at a checkpoint in each village square had punched a stopwatch to measure the next leg of your trip.

At stop signs you half expected a crew to swarm all over the car, adding fuel, washing the windshield and handing you a drink. In the background of this fantastic scene, of course, was a dark-eyed man in a baggy suit and a hat, watching the proceedings with an air of detached authority, knowing he had done all he could and things were proceeding as they should.

The Ferrari, perhaps any Ferrari, is a car of high drama in appearance and sound. It's impossible to drive one and separate reality from that rich overlay of myth, racing history, famous drivers, Italian roads and great endurance contests. The heritage goes with the car.

All through the trip I had a secret desire to rise very early one morning while Chris slept, walk to the outskirts of town with a bucket of whitewash and daub VIVA DINO on an old stone wall in large, runny letters. That, or paint over a milestone to read BRESCIA 98 KM.

Lofty goals for a guy who barely has the energy to comb his hair in the morning.

In California we took Highway 102, one of the great sports car roads of all time, past Mono Lake and then turned south to Highway 395, down through the Owens Valley. At Lone Pine we saw our first motorcycle of the trip and passed a Frosty Root Beer stand, where a chilled drink looked good for the first time in six days. We rolled the windows down and at China Lake took our jackets off and put them in the trunk.

At Four Corners we turned the heater off for good and opened the vents for the first time. It didn't seem to make any difference. Were the vents open? Had the heater ever worked? We couldn't tell.

I played with the stiff, ineffectual vent levers and it occurred to me that the Ferrari is a car that's the sum of its parts. No single piece of the car has the handcrafted appeal of, say, an old Jaguar or MG where you pick out chromed knick-knacks and polished castings as something special. The Ferrari's hardware—window winders, handles, instruments, latches and switches—are for the most part quite ordinary, or even substandard in some cases. Where the Ferrari's money has gone is into a tough, beautiful engine, a rugged frame and a firm-but-compliant suspension that is the product of lessons learned from thousands of miles of road racing on real roads. There, and into the sleek, lovely sheet metal covering the whole marvelously competent package. There are few mantlepiece conversation pieces to be saved off the chassis of a wrecked Ferrari. It has its foremost value and greatest beauty as a single working piece.

When we pulled into my driveway, the Dino had worked flawlessly for six days, not counting the required alignment and wheel balancing. It averaged 21 mpg over 2900 miles of fast to very fast driving. It started at 10 below zero, pushed on through drifted snow, took every pothole the road dished out and handled mountain roads with exceptional balance and agility. Its heating system built great character.

Furthermore, the car looked good in my driveway. The day after the trip I got out the hose and a pail of hot soapy water to wash the long week of road grime off the car. Running a sponge over the curve of the front fenderline I began to wonder, idly, how much money a man could raise if he sold everything he owned, except for his old British sports cars.

RETURN OF THE BUGEYE, PART I

Finding the right car wasn't easy; keeping it was harder still

BY PETER EGAN

PHOTOS BY THE AUTHOR, LEE HEGGELUND, JOHN JAEGER & LINDA ROTH

New Year's Day, 1974: One classic follows another.

Restored formerly-blue Sprite sold to man in Milwaukee. Car was stolen a week later, never recovered.

Tying it down; stripped-out Sprites weigh more than you think.

Still orange, 1st time off the trailer, driver's school, Blackhawk Farms, '74.

1st racing car morning after the all-night drive from Michigan.

New driver's suit, new car, before 1st race. Oiled clutch, spun pilot bushing, but car finished anyway.

CERTAIN CARS, LIKE the refrain of a movie sound track, have a way of turning up and then fading from our lives, only to reappear during a later part of the story. The musical treatment during different parts of the film might be upbeat or somber to fit the required mood, but the song is always the same. In my own case, the mechanical reprise happens to be a funny little English sports car for which I've always been too tall, a machine produced for only four years in the late Fifties and early Sixties. I've owned six of these things, one of them twice, proving that if History teaches us anything, we're usually sleeping in class. Or sketching cars in the margins of the textbook.

This is a story about the funny little English sports car I owned twice, and it starts in 1973.

You remember 1973. No? I didn't either, so I looked it up. That was the year of the Yom Kippur War and the defeat of Bobby Riggs at the hands of Billie Jean King in the Battle of the Sexes. It was a dismal year for music, so I won't mention any of the songs that were popular lest they stick in our heads and irritate us all day, but *Half Breed* by Cher was right up there with *Tie a Yellow Ribbon* by Tony Orlando and Dawn. Sorry. Nixon was up to his five o'clock shadow in hot water over the Tapes, and there were no more U.S troops in Vietnam.

On a more personal note, it was also the year I suddenly rose from the Turn 6 bleachers in the middle of the H Production race at Elkhart Lake, Wisconsin, vowed quietly that I would never again attend a sports car race until I'd become a driver rather than a spectator and walked down Fireman's Hill to a racer's

BRG at last! Orange car right after proper paint job.

Kitty, back porch, 1974.

At speed, more or less, Blackhawk Farms, 1975.

BUGEYE

supply van where I bought a complete Nomex driver's outfit for $99.95, including underwear, hood and socks.

This was a bold move for a guy who didn't own a racing car and was then dragging down, easily, two figures per week as an assistant rain gutter installer and backyard mechanic. Nevertheless, I drove home from Elkhart Lake, hung the driver's suit on the back of the closet door where I couldn't miss seeing it every day and began to look for a racing car. Like so many people before and after me, I decided that the cheapest and most satisfying car for the launching of a racing career was an H Production Austin-Healey Mark I Sprite.

A Bugeye.

Why a Sprite? Many reasons. First, I liked the looks of the early Sprites, especially in racing trim. The sheet metal, though simple, came together in subtle flowing curves at the rear, and I found the face formed by the headlights and grille (which the English think is a frog and we think is a bug) to be engaging and friendly rather than ugly. When the Sprite was introduced in 1958, some automotive critics ridiculed its appearance and later applauded Austin-Healey's switch to the more rational but much less interesting Sprite Mark II. To my eye, however, the first Sprite was the best one, built the way God and Donald Healey intended, and I was disappointed when the style changed.

The Bugeye was such a pure little shell of a sports car right from the factory that very little extraneous junk had to be removed for competition; no door handles, radio consoles or windows to pull out and throw away. When you peered into a Bugeye all you saw were two barrel-backed seats, a shift lever, a couple of instruments and a steering wheel. Rubber floor mats were standard, carpets extra. It started out closer in spirit to a pure racing car than any other production sports car I knew of, except perhaps a Lotus Seven or a Porsche Speedster.

Beyond this private storm of esthetics, there were other good reasons for racing a Sprite. Structurally, the car had a wonderfully strong, unified steel body and frame welded up in rugged box sections, and the entire one-piece bonnet assembly could be made to tilt forward or lift off for easy access to the engine bay. And once you got there, the Austin drivetrain and live rear axle were dead simple to repair, like a child's textbook diagram of How Cars Work.

At that time the basic car, in both Sprite and MG Midget guise, had been in production for 16 years, so there were tons of spare parts lying around in junkyards—and backyards. Shock absorbers, disc brakes, A-arms, transmissions, wheels, differentials, etc, could all be lifted straight out of a 1973 Midget and lobbed right into a 1959 racing Sprite.

Only the original 948-cc A-series engine had to be left in place, rather than the later and stronger 1098- and 1275-cc engines, to comply with H Production displacement rules. This little powerplant had three main bearings, two 1 1/8-in. SU carbs and a long stroke, and could be lifted out of the car, bonnet removed, by two casual people or one desperate person. The engine responded well to tuning and could (I later learned) actually be revved to 9000 rpm without blowing up. Sometimes.

New parts were also cheap and readily available. The Sprite used the smallest and least expensive of everything in the J.C. Whitney catalog. If the blurb at the top of the page said, "Piston sets as low as $39.95," those were usually Sprite pistons. "Exhaust valves low as $2.99" were Sprite valves, of course, and you could buy an entire exhaust system for just $13.00. (I now have an old Jaguar E-Type, and all the parts are at exactly the opposite end of the Whitney spectrum.)

Okay, so a Sprite was the car for me. But where to find one?

Summer ended and half the winter slipped by while I searched for the right car. There were plenty of used Bugeye racing cars for sale in the *Autoweek* classifieds, but I was determined to start with a street car and build my own racer from the ground up. This complicated things, because there weren't a lot of sound Bugeyes running around on the streets. By January, I'd looked at half a dozen prospective Sprites, all of which proved on close examination to be too wrecked, too rusted or too deeply frozen in snowbanks to meet my high standards. By insisting that the car have some suggestion of where the floor had been, I severely limited my choices in the salted Midwest.

I was lamenting this dearth of good cars over dinner one night with a group of people that included my good friend and fellow soon-to-be racer John Jaeger. We were all sitting around dressed in work boots and old flannel shirts that looked like something Neil Young might have thrown away, eating spaghetti and drinking quite a bit of Boone's Farm and Annie Green Springs, as was the fashion then. (Health spas, PCs and money man-

agement were still six or seven years in America's future at this point, so we didn't have anything better to do.)

Toward the end of dinner, John suddenly held one inspirational finger in the air like a saint on an Italian mass card. "I've got it," he said. "I know where you can find a perfect rust-free Sprite, completely disassembled. I should have thought of it earlier."

"Where?" I asked, predictably.

"Hancock, Wisconsin."

I gave him one of my well rehearsed blank looks, recently perfected in college math. "Where?"

"Hancock. Little town south of Plainfield, not too far from Vesper and Arpin . . . It's up north."

"I can picture it."

"An old high school friend of mine named Mike Nieman lives there. When we turned 16, Mike and I both bought Sprites and drove them around all summer. Mine was a Mark II and Mike's was a Bugeye. In the autumn, Mike parked the car in his parents' basement garage and stripped it down to the bare chassis. He painted it dark red over the old powder blue, cleaned and labeled all the parts and put them in boxes. That's as far as he got. He never put the car back together. When Mike got married he moved the Sprite into the basement of his new house. As far as I know, it's still there, sitting on blocks. No rust, no dents. The car was perfect."

Being a gentleman, I didn't force John to call his old friend Mike Nieman immediately. I waited until he'd finished his drink.

We went to get the car on the coldest day of the winter, three semi-employed friends and I, driving 100 miles north on Highway 51, towing a borrowed trailer with a borrowed Rambler Classic. Despite the cold, this car managed to overheat just a few miles from town. We stopped at a farmhouse to ask for some water, and the woman who came to the door thought our request was some kind of practical joke. "No car could overheat in this weather," she said suspiciously.

"It's a Rambler, Ma'am," I said, shivering on the porch.

She nodded and went to get me a pitcher of water. We added two quarts of water and continued north. The Rambler had a hole in the firewall where the rusted-out heater had been, so we froze our feet. I'm not speaking figuratively here. We froze our feet.

The overhead door moved reluctantly in its tracks, sliding upward with a great deal of frozen creaking and groaning.

BUGEYE

Low winter sunlight slanted into the basement garage, and Mike Nieman said, "There it is."

There it was, all right, that eternally cheerful face grinning from beneath an old wool army blanket. Five of us moved into the garage, lifted the blanket and milled around the car, stamping our feet and breathing steam like horses crowded into a barn. It was 15 below zero outside and the cold air snaked into the basement, lifting small trails of vapor off the warmer garage floor.

Sitting high off the ground on two sawhorses was the most perfect Sprite I'd ever seen, this side of a showroom. The chassis was immaculately clean and rust-free, painted a dark claret red and stripped to an empty shell. Only the rear axle and a few suspension pieces had been bolted back together. I don't know how a car smiles, I thought, after sitting in a dark basement for seven years with its wheels off, engine block on a shelf, seats in the rafters and valve gear stacked in the bottom of the oil pan. Maybe the Sprite was pleased to be stored in a warm, dry basement while so many of its kind were out there on the road, losing their floor pans to salt. While other cars were being consumed by their owners, this one was laid on its side in the cellar, like a nice bottle of wine, gathering dust and character.

I wrote Mike a check for $500, and we all hefted the chassis onto the trailer in a sort of shuffling hernia drill and cursing contest. We filled the hollow car with boxes and coffee cans and baby food jars full of parts, slid the block into the trunk of the Rambler and drove off into the cold. No one who was there has ever completely forgiven me for the frozen feet or the weight of the car.

Even before we'd returned home, I knew the Sprite would never be a racing car. It was too perfect, too complete. In the parts inventory were four unblemished A-H hubcaps, all the original badges and emblems, a flawless pair of bucket seats and a speedometer with 22,000 original miles on it. A twist of fate had preserved at least one Bugeye in all of the Midwest, and it was not my place to go cutting into that unspoiled sheet metal to make fender flares and install a rollbar, or otherwise mess up such a pristine little road car. I was obliged to play my part properly in this lost dog story. The only right thing to do was reassemble the Sprite and sell it off to a good home.

That job took exactly one winter. In the years of storage, many small pieces of the car had been lost, falling through mystical cracks, as only parts in coffee cans and cardboard boxes can. If four special screws were needed to bolt the windshield to the body, then there were only three in the coffee can. Of nine cyl-

inder head studs, only eight could be found, and so on. Luckily I lived down the street from a hotbed of Anglophilia called Foreign Car Specialists where a parts man named Scott Coffrin carried an exploded view of the Sprite, complete with parts numbers, in his head. He also had bins of odd British screws and fittings (as if there were any other kind) and knew exactly where they were.

The project got another shot in the arm when I found a Bugeye parts car rotting in a backyard only three blocks from our house. This car was light blue and had its engine scattered around the trunk in a sort of shotgun blast of parts. The crankshaft was sitting in the driver's seat and the cylinder head was lying in 2 in. of rainwater and wet leaves on the floor. I paid the owner $300, pumped up the tires and pushed the car back to my garage singlehanded (try that with your fancy 4.5-liter Blower Bentley). This car turned out to be deceptively sound and complete under all the grunge, so I later restored and sold it, but that's another story. Back to the Bugeye at hand:

I reassembled the chassis by day, and at night attended engine building classes at the technical and vocational school. They had a hot tank, a valve grinder, a bead blaster and central heating, so I kept taking the course over and over again. (I think the fifth time I signed up for Introduction to Engine Rebuilding, they put up a NO LOITERING sign in the registration office. Winos had their bus stations when wind was cold; I had the Madison Area Technical and Vocational Schools.) I installed new bearings, lapped the valves, honed the cylinders and re-ringed the pistons. The crankshaft showed no wear, amazingly, and got by without grinding. One piece at a time, the car went back together, a garage-sized jigsaw puzzle of bushings, bearings, brake parts, wires and lights.

One particularly fine spring day the Bugeye was finally wheeled out of the garage. The new engine cranked over slowly and then fired up in a blue cloud of assembly oil, shaking and jittering like crazy—a problem quickly tamed with the application of an SU gland nut wrench, a screwdriver and a piece of rubber hose, which is not as off-color an operation as it sounds. Carbs adjusted, the engine ran fine, settling into that soft, hollow idle that always makes me think of wartime Jeeps at RAF airfields.

My wife Barbara and I took the Sprite for an all-day drive in the country. Though the Sprite was 12 years old, it was mechanically the newest car I'd ever

owned. The steering was quick and precise, the body rattle-free, and the engine revved with a wonderful small-bore snarl. The only other sports cars I'd driven for any length of time were my two old Triumph TR3s. The Triumph was a more stately car to drive, and much faster, but it lacked the Bugeye's tight, nimble handling and seemed almost truckish by comparison.

The Sprite could be pitched and drifted through corners with abandon, and, as with early Volkswagens, you could race it around town without anyone knowing. The car was an absolute gas to drive, and felt the way magazine writers always told me true sports cars should feel but never did. We motored along all day on winding county roads through the hill country of southwestern Wisconsin, top down, Barb and I taking turns at the wheel.

On the way back to our house, early in the evening, Barb said, "Isn't there any way we can keep this car? We'll never find another one like it." I looked at her and shook my head.

"Racing," I said. "We're going racing."

After a couple more days of shakedown driving, I turned the keys over to a friend named Pete Koerwitz, who had been waiting patiently to buy the Sprite the minute it was reassembled and ready to go.

Pete gave me a check for $1600. This large chunk of money, in its entirety, was immediately turned over to a man in Michigan who then allowed me to haul away his flaming orange race-prepared H Production Bugeye. (I later painted it British Racing Green, of course. Given a chance, I paint nearly everything British Racing Green. Around my Irish father, however, this color is always referred to as Kelly Green.)

The Michigan deal included a trailer and three spare engine blocks with connecting rods sticking through their sides, as a sort of preview to the wonderful world of racing and engine tuning. I had a racing car at last, and the Nomex driver's suit came down off the back of the closet door. After that, there was no looking back. There was looking broke and there was looking perplexed, but there was no looking back.

I never really lost track of that first Bugeye, keeping tabs on it the way you do old friends from high school, with a question here, a chance meeting or an overheard conversation there. I saw the car on the road fairly often. Pete Koerwitz and his wife Cheryl commuted from their country home into the city for three summers with the rebuilt, wine-red Sprite. They had it repainted an unfortunate bright parrot yellow that neither of them liked, but it was too late to do anything about it; a simple case of misreading the typical 1-inch-square paint chip at the body shop.

When the gearbox became noisy, as Mark I Sprite gearboxes are wont to do, they parked the car and shifted the commuting duties to a new Ford pickup. The Bugeye languished in a dark corner of their garage for another three years, gathering dust and cat footprints.

About the time Barbara and I moved to California (1980), they sold it to another friend of ours named Hajime Nakayama, a Japanese university student. Haji, as he was nicknamed, was a knowledgeable British car nut and math wizard who had waist-length hair and, at the time, looked remarkably like Geronimo. Only his habit of smoking with an FDR-style cigarette holder brought him into modern focus. Haji drove the car one summer, then parked it in a rented storage unit and moved to San Francisco with his other Bugeye and his Honda 600S. Haji planned to go back and get the car sometime, but he never did.

I kept a picture of the Bugeye up on my wall at the office. Every so often I would look up from my typewriter (this implies that I sometimes look down at my typewriter) and consider the car's fate. "There it is again," I thought, "locked in another dark garage." The Bugeye seemed destined to make brief forays into the sunlight, then to be put away and forgotten. It always suffered from being fun rather than essential.

I knew exactly where the Sprite was kept, in a dismal row of earth-tone rental storage compartments, each one numbered, all of them filled with the unwanted furniture, bicycles and barbeque grills of people who were temporarily rootless. It seemed like the wrong place for an old English sports car, like making Holmes and Watson live in a condo.

The car gnawed at me somewhere deep in that misplaced sense of animism that fosters sympathy for cars as though

Sprite in good company, party at Chapman farm, spring '74.

they were conscious living things. Druids, I'm told, had the same problem with oak trees, while normal people spent these emotions on cats, dogs and canaries. Something had to be done.

Then there came an afternoon last spring when the wind was blowing just right, carrying that damp-earth smell that makes people run out and buy convertibles, motorcycles and biplanes on credit; the kind of day upon which irrational behavior is defensible in a court of law, provided the jury leaves the courtroom windows open and the lilacs are blooming. It was a dangerous day, in other words, to be looking at pictures of a favorite old sports car. Especially a sports car locked in a garage in a part of the country where I knew by instinct and the angle of the sun that the last patches of snow were just now melting on the north sides of buildings and trees. I picked up the phone and called Haji. Long distance.

A secretary in the bank where he works as some sort of roving computer expert tracked him down in the hallway and brought him to the phone.

Yes, Haji said, he just happened to be thinking of selling the Sprite because he didn't know how or when he would ever bring it out to California. "It'll need a lot of work before you drive it anywhere," he said. "It needs brakes, a battery, a new transmission, a clutch job

and bodywork. Also, the side curtains were stolen." I would, he added, find the car rustier than I remembered, but not unrepairable. The tires were new—Michelin radials—and he'd had the front kingpins rebuilt before the car was stored. He would sell the Sprite for $1350, which was exactly what he had tied up in it. No hurry on the money. It would do his heart good, he said, to see the car fixed up and on the road again.

When I got home that evening, I pulled one of my favorite books off the shelf, *More Healeys* by Geoffrey Healey. I sat for some time, turning pages and looking at the pictures: Moss, Sprinzel and McLaren drifting their Bugeyes around the flat airport turns at Sebring, Tommy Wisdom flashing through the green Sicilian countryside in the 1959 Targa Florio in his bright red Sprite, and Donald Healey himself, standing proudly by his creation at the press introduction at Monte Carlo in 1958.

A new Sprite cost less than $1800 in those days, and it always amazed me that such a small, inexpensive car could have stirred up so much fun and excitement, or have generated so much history in such a short time. Cheap though it may have been, the Sprite also had a special, friendly kind of exclusivity; it excluded people who took automobiles, and maybe themselves, just a little bit too seriously. In all the pages of Geoff Healey's book, I couldn't find a single photo of anyone frowning. The Bugeye was a type of machine all too absent now, a car upon which people beamed.

I laid the book aside and said casually to Barb, "Guess how we might possibly like to spend our summer vacation this year."

She lowered the evening paper just enough to throw me a level, studied gaze over the top. Then she folded the paper neatly and put it on her lap. "I thought we were going to buy an old Piper Cub," she said, "and take it on a leisurely trip to Wisconsin to visit our old friends . . . fly slow, you said, see the small towns and the green Midwestern countryside from low altitude . . ."

"Close," I said. "Very close."

RETURN OF THE BUGEYE, PART II

Bringing it all back home

BY PETER EGAN
PHOTOS BY THE AUTHOR
MONTAGE BY BRIAN BLADES

T HIS IS WHAT it's finally come down to, I thought. Flying halfway across the country by DC-9 in just three hours so you can buy an elderly little sports car with pistons the size of Contadina tomato paste cans and spend a week driving it back across the country. Sort of like taking an air-conditioned Cadillac to see a man about a horse. So many of our adventures now come from backsliding through technology. The frontiers are gone and everything is easy, so you do it the hard way. When there are no more outlaws, you shoot yourself in the foot. On purpose.

The hard way, in this case, promised to be slightly more fun than the average self-inflicted wound, provided certain aged fragments of cast iron, steel, bronze and cotton-covered wire felt like cooperating. My wife Barbara and I were flying from California to Wisconsin to collect one of our favorite old cars and bring it home. The car was a 1961 Austin-Healey Bugeye Sprite, which we had restored and reluctantly sold in 1974. Like parents who give children up for adoption in times of economic stress, we hadn't forgotten.

When we sold the car more than a decade ago it was almost perfect; rebuilt engine, no body rust, new top and nice red paint. Two owners and 45,000 miles of backroad commuting had changed all that. Now the car had a slipping clutch, 2nd gear gone from the transmission (with others anxious to follow), a noisy timing chain and low oil pressure. It was also blessed with a $60 parrot-yellow paint job with overspray on the tires and seats, as well as a top that looked like "quite a lot of rain could fall in there if it wanted to."

That quote came from my old friend and former employer, Chris Beebe, who had graciously towed the ailing car out of a storage barn and back to Foreign Car Specialists, his repair shop in Madison, Wisconsin, for inspection. "You'll

BUGEYE

need to send for a new or rebuilt transmission," he had told me over the phone. "We have most of the other parts here at the shop. I know a good body man named Steve Linn who can fix the rust and paint the car before you get here. You don't want to cross the country in a Bugeye painted this shade of yellow. It reflects badly on your friends."

"Anything else?"

"Bring side curtains. It doesn't have any. I've got a good used top, but I can't find any side curtains. By the way, what color do you want us to paint the car?"

"Surprise me," I said. Chris knew I had never willingly painted a car anything but green.

"See you in two weeks."

I found a British auto supply place that sold rebuilt Sprite gearboxes and had them truck one off to Madison, then located a well-used set of side curtains at the Austin-Healey Store, a restoration and parts business in Reseda, California. Old Sprite side curtains, it's worth mentioning, are an exact machine-fit in the Size Large American Tourister suitcase, and they leave black aluminum oxide stains on your clean undershorts.

Barb and I flew to Madison during the first week in July.

I spent the first week of summer vacation under the Sprite. While Steve Linn finished rubbing out the very nice BRG lacquer he'd used on the car, I pulled the engine, replaced the timing chain and put the engine back in with a new clutch and transmission attached. In three years of storage, the Sprite's brakes had frozen and the brake fluid had turned into a dark, espresso-like substance, so I rebuilt the wheel cylinders and the master.

When we finally started the engine, it had a strange noise in the bottom end and only 35 psi of oil pressure, so we pulled the oil pan. The main and rod bearings looked all right, but the distributor drive appeared to have been chewed by rats, so we installed a new (used) one and all the noises went away. We attributed the low oil pressure to cam bearing wear, a common problem on early 948-cc Sprite engines. Installing new bearings was a major disassembly and machine-shop operation for which we had no time. It was Saturday night and we had one week to get back to California.

"Cam Bearing Rebuild in a Can," I explained to Barb, dumping a container of STP into the valve cover. Our oil pressure shot up to 38 psi.

We left Madison on a Sunday July

morning and aimed our grinning grille southwest, toward Iowa and Missouri. The southwestern quarter of Wisconsin is my favorite part of the state. Hill country, carved up by rivers and streams, it's more Ozark-like than typically Midwestern, with towns that are small enclaves of old-worldliness, Swiss, Welsh and otherwise. The winding country roads follow the river valleys, and it's a good place to try out a Sprite when you haven't driven one in years.

All the old sensations were there. The first thing you notice about a Sprite is that the steering is quick. Very quick. People who are unprepared occasionally run over curbs and hit stop signs while trying to make simple right-hand turns on city streets. You learn to sneeze *without* twitching the steering wheel, much the way a good waiter learns to trip without spilling his tray of drinks.

Your next impression is that the car is very short on the outside while being relatively roomy on the inside. I am 6 ft 1 in. and have no trouble getting comfortable in a Sprite. The car has a deep, horizontal footwell for good leg room, and the doors are hollow shells, so there's plenty of elbow space. Later Sprites and Midgets got padded out and claustrophobic with wind-up windows, federalized safety instrument panels and collapsible steering columns, but the early ones fit almost anybody.

Beyond that, the Sprite felt light and jaunty and just plain fun to drive again. It accelerated with a nice ripping sound from the exhaust, stopped much better than cars with all-around drum brakes are supposed to, and had a very short clutch throw and a shift pattern to match. The gearbox I'd installed was a late-model close-ratio version, so it didn't have the gaping ratio holes of the original Bugeye box.

All the car really lacked (other than about 1000 lb of Modern Features and road-hugging weight) was a 5th gear. Sixty-five miles an hour equaled 4200 rpm. All day. There was a vibration period in the engine right around that speed, and we found cruising to be smoothest just under 60 mph, at around 3800 rpm. The gearing was fine on English-style lanes, but a little too hectic on the long American straights. I looked at the tach and the oil pressure gauge and tried not to think about the Mojave Desert. Or Kansas.

We stuck to the minor roads, those stray, low-voltage circuits in the national wiring diagram, drawn by Rand McNally in faint gray, and jogged

through Iowa and into Missouri through small towns with large John Deere dealerships. Between farmers gathering hay and town people mowing lawns, there was a good summer smell of dust and fresh cut grass in the air. Late in the afternoon we saw a farm kid shooting buckets at a hoop nailed to the side of the barn. The sight of our car caught him in the middle of a free throw and he missed the shot. The ball bounced in the barnyard with small explosions of dust, and in the mirror I saw the boy wave. I tapped the Sprite horn and it honked a single weak bleat, then quit working entirely. Barb looked at me and I shrugged.

"Short," I said.

On these rural roads we noticed that other drivers tended to follow the Bugeye very closely for long distances, even when there was plenty of opportunity to pass. I wasn't sure if this came from natural curiosity or merely a desire to loom. After dark it was worse, as headlights in the rear window of a Bugeye began to look very large indeed, especially headlights of a tall pickup truck crowding your rear bumper. The Sprite's mirror and rear window are so small you can usually see only one headlight at a time and it's all out of proportion, like having a giant ape or a praying mantis peer in your 5th-floor office window. It makes you uncomfortable.

Leaving the city limits of Milan, Missouri, I hit the high beam switch with my foot and all the lights went out on the car. There was a sizzling sound from beneath the instrument panel and a smell of burnt wire wafted through the cockpit. I clicked the switch back to low beam and all the lights came on again. The sizzling and smoke subsided.

"Short," I explained to Barb. "I think we should avoid using high beams from now on."

We found a nice little motel (FREE COFFEE & DONUT BREAKFAST) in Cameron, Missouri and unloaded our bags. The Sprite doesn't have a trunk lid, so everything including the spare tire gets stashed behind the seats in the relatively cavernous rear body section. It takes a flashlight and a certain amount of tunnel-rat tolerance for tight places to retrieve your luggage, but the trunk holds an amazing amount of gear.

In the morning Barb and I went to the motel lobby for our free coffee and donut breakfast. As we poured our coffee, the lady behind the counter said, "Careful. I just took those donuts out of the freezer, so you'll want to dunk 'em."

I picked up a donut, which had about

the same density and heft as a snub-nosed revolver. The lady watched as I dunked my first donut and bit into it. "Is it thawed?" she asked.

"Yeah," I said, "but now my coffee's cold."

We crossed the Missouri River at St Joseph and drove into Kansas. Near Denton we drove down a straight, rolling stretch of road and watched a pair of crop-dusters doing a slow ballet, criss-crossing the highway. They shut off their sprayers while crossing the road, but a fine mist and heavy chemical smell still drifted over us. "Is it possible," I asked Barb, who is the medical expert in our household, "that a chemical can rain down leaving locusts lying on their sides kicking and gasping, while we drive through unaffected?"

"I think with us," she said, "it just takes longer to work."

Except for the chemical spray and the tendency of farmers to take short naps at stop signs, I've always enjoyed driving through Kansas. The farms have an established, stately feel to them, and I like the towns. As centers of rural commerce, they have a great self-sufficiency. You have the feeling that the continent could fall into the ocean in a hundred-mile circumference from town and life would go pleasantly on. In Los Angeles I always worry that a truckload of ball bearings or chickens will tip over on the 405 Freeway and, in the ensuing traffic jam, we'll all starve to death.

For a car like the Sprite, unfortunately, the Great Plains are not the best environment, being essentially one long broomstick test. If pistons and valves could think, they'd probably go mad with heat and boredom in a place like Kansas. It's been my experience that British things (especially motorcycles) don't like to drone along at one unvaried rpm, so in crossing the plains I frequently lifted off the accelerator, just to suck a little extra oil past the rings and guides and to let the engine know I was thinking about it. I didn't know if this really helped, but I figured superstition and ignorance had kept me alive this long, so why meddle with a good thing.

The humidity in central Kansas, by late afternoon, was just a few percent away from bathwater, so we stopped at a store and got some big slurpy drinks with red stuff poured over ice. Back on the road we were reminded that the Sprite has stiff springs and is not a good drinking car. Bumps come as sharp raps to the seat bottom and translate immediately into nosefuls of hot coffee and fizzy soft drinks. Or red stuff and ice. It's hard on the shirt, too. After five miles we looked like extras in a Sam Peckinpah movie. You learn to time your sips between punches, drinking with one eye over the rim of your cup, watching the road.

BUGEYE

We followed Highway 24 nearly all the way across west Kansas, through places like Glasco, Alton and Nicodemus, keeping the south fork of the Solomon River just to our left. Early in the evening we stopped at the Western Hills Restaurant in Hill City. Barb had chicken fried steak and I took a chance on the fried oysters (Hill City is a long truck drive from most of your well-known oceans). The place was full of ranchers.

Two serious young men, maybe 18 years old, in tall cowboy hats and vests sat down at the table next to ours. When the waitress came they both looked up at her and said, "Special." The Special, when it came, was one of those blue plate deals that appear to include a side of pork, dressing, six or seven kinds of yams and potatoes and a couple loaves of bread, all swimming in dark gravy. The two cowboys worked their way through these mountains of food, hats on all the while, without saying a word. When dinner was over, each of them took out a tin of Copenhagen and carefully placed a pinch between cheek and gum. They sat back in their chairs and stared out the window for about 10 minutes, checked their pocket watches at exactly the same time, nodded at each other, paid their bill silently and clumped out of the room in well worn cowboy boots.

"Off to their tennis lesson," I said to Barb. "So much for McLuhan and the Global Village."

"Pretty good," Barb said, "getting through a whole meal with just one word."

"Special," I said tersely, practicing my Western.

Outside the restaurant, our Sprite looked really menacing parked in a solid row of pickup trucks. Mr Peepers Goes to the Rodeo. Barb said it looked like a character in a Dell Golden Book entitled either *The Littlest Sports Car* or *The Sports Car that Could.*

Some of the towns on Highway 24 are fairly large, but you have to veer toward the Interstate these days to find a motel in most parts of the country. In California and New England industrious entrepreneurs may be refurbishing old inns and hotels, re-oiling dark woodwork and lavishing potted plants upon them, but in Middle America the grand old downtown hotels are either havens for old men with emphysema or they are nailed over with plywood sheet.

We drove down Highway 83 and found a motel in Oakley, Kansas, a nice

little town just off Interstate 70. Early the next morning we had breakfast in a place called The Little Cafe, on Main Street. We got out the road atlas to check our progress and found that we were almost out of Kansas. I looked out the cafe window at our Sprite and then back at the map. The two of them did not relate.

Mapmakers have it all wrong, I thought. They give you the same size map no matter what kind of car you're driving. They need to come up with a new formula based on displacement, wind noise, gear ratios and spring stiffness. These little one-page maps of Kansas were fine for traveling by Cadillac or Lincoln Continental, but they were far too small for a Sprite trip. They were all out of scale and did not provide fair warning as to the size of a state. The Sprite needed a larger Kansas map, something about the size of the American flag on the Pentagon.

We crossed into Colorado at noon. Eastern Colorado is that part of the state where the land saved up its strength for the Rockies, which rise out of the plains just to the west. We took Highways 40 and 94 toward Colorado Springs, a couple hundred miles of desolate, straight road and isolated towns. "Flat," I said, sweeping my hand over the landscape. I was beginning to like the Western school of speech. It saved a lot of shouting over the wind noise and valve clatter.

A car like the Sprite feels like a very small thing crossing the prairie. The mind's eye leaves the car and watches it from far overhead, a small green speck traveling across the brown grassland. It's dwarfed by the size of the country, like a distant Piper Cub against a sky full of thunderheads. There were no thunderheads to worry about, but we had that tiny engine, pulling us across miles of open space. All with the original crank, original pistons, rods and camshaft, nothing bored out or ground undersize. Nothing new in 23 years except bearings, rings and a timing chain—and a distributor drive. A little old engine with a long stroke, non-free floating wristpins and Austin written on the valve cover.

We sped up to 70 mph to pass a fertilizer truck that was throwing rocks and dust and other things best analyzed by a state laboratory, and I tried to remember exactly how I'd built the Sprite engine in technical school, 12 years ago. Did I torque down all the flywheel bolts properly? Use Loctite on them? Red or Blue Loctite? I looked at our 35 psi (hot) of oil pressure and tried to picture what was going on in the engine. Visions of abundant oil flow are important for peace of mind on old British cars. I worry about British cars the way I worry about things I've built myself, even when I haven't, and it's worse when I have.

We stopped at a gas station in Kit Carson, Colorado, and I added another can of STP. "If this engine makes it to California," I told Barb, "I'm not going to rebuild it again. I'm going to have it bronzed and set it on top of the TV."

The Rockies appeared above Colorado Springs late in the afternoon, with thunderstorms boiling around their peaks, and we drove toward the mountains with wonder and misgivings, like Dorothy and the gang approaching Oz. Planning an assault on the mountains in the morning, we stopped at a speed shop at the edge of the city and filled our tank with high octane leaded racing gas ($2.00/gal.), which I hoped would reduce pinging and help the Sprite survive the long climb to the Great Divide.

Filling the tank, I found our mileage across east Colorado had been the best of the trip, at 39 mpg. Our average up to that point had been about 35 mpg. Years ago, when I'd first rebuilt the engine, the Sprite's mileage had never dropped below 40-plus mpg, making it a great companion to our 11-mpg 1966 Ford wagon. We've become accustomed to cars getting great mileage now, with 5-speed Honda Civics and the like, but in the early Seventies 40-plus mpg was motorcycle territory; a spectacular figure for anything on four wheels. It still isn't bad for a car with two valves per cylinder, a 4-speed transmission with a 3.22 rear end and no microchips.

We stayed the night in Colorado Springs with Barb's aunt and uncle, Howard and Pauline Rumsey. Howard is a born craftsman and handyman, and it bothered him that the Sprite had a couple of Phillips screws missing from the cockpit trim, so he dumped his lifetime nut and bolt collection on the floor of his garage, and we found screws of exactly the right size and shape. We were treated to a steak dinner, a sunset view of the Garden of the Gods and Pikes Peak from the Rumsey's patio, and half an hour in their heated spa. Later, Howard helped me wash the Sprite. I'm told that not everyone has relatives like this.

The early part of the drive into the mountains was beautiful in the morning and the Sprite ran fine, sweeping up through the mountains easily in 3rd

gear. Near Wilkerson Pass, however, the usual collection of motorhomes clogged the works, and continuous 2nd-gear driving boiled the water in our radiator. We pulled over a lot to enjoy the scenery while the car cooled. Windex, sprayed on the radiator, seemed to bring the temperature down faster, or at least we believed it did (superstition, ignorance, etc). In my mind I designed a sophisticated Windex injection system with a reservoir tank and a row of spray nozzles in front of the radiator, a device that would revolutionize mountain travel for Sprite owners everywhere.

Uphill progress was slow, but downhill travel was another story. The engine cooled down, our passing potential soared and on winding roads the Sprite made mincemeat out of most cars on the highway. We could have used a front anti-roll bar, as a stock Bugeye has a bit too much body roll, but its handling was still astoundingly better than the average American sedan, and it had no trouble staying with the more aggressively driven econoboxes. Generally, the Sprite had the same problems and advantages on the highway as it had on the race track—it ate up sedans in the corners and got blown away on the straights.

As we drove along the forested canyons of the Gunnison River, Barb said it

should be a national law that all cars have bullet-shaped headlight housings. The curves of this shape are such that clouds, trees, mountain peaks and overhead bridges fan out and then converge and quickly disappear in beautiful reflected patterns; a constant entertainment. I was glad the Sprite hadn't come with retractable headlights the way Donald Healey originally intended. Without the headlights we might have wanted a radio.

After a lunch at our favorite Mexican restaurant in the west, the Stockmen's Cafe in Montrose, we turned south along the Animas River on our favorite mountain road in the west, Highway 550 from Ouray to the restored mountain mining town of Durango, and stayed at our favorite hotel in Durango, the lovely old Strator. The only room they had available, due to a cancellation, was their best room, the Presidential Suite, so we took it. Just another tough day on the road.

The next day we drove into a mountain storm, the first rain of the trip, and dug the side curtains out of the trunk. Up until this point we had installed the side curtains only when the car was parked at night, in case of rain. These little clip-in windows fit very snugly on

BUGEYE

the Sprite, making it one of those rare British sports cars of the Fifties that will actually keep you warm and dry when the weather is not. We removed the side curtains later in the day and discovered the airflow is so clean around the Sprite's windshield and top that very little rain blows in even with the sides open. We stored the side curtains in the trunk and left them there through New Mexico and the rest of the trip.

As we crossed into Arizona, the speedometer needle began swinging back and forth like a metronome, keeping time to some snappy song beyond the range of our hearing, possibly an early Xavier Cugat number with maracas and conga drums. Half an hour later the needle laid itself to rest at zero and stirred no more. About five miles down the road we drove through a large puddle and the fuel gauge shot over to Full and sat there quivering and twitching. Then it abruptly dropped to Empty.

"Electrocuted," I explained. "Must be a short."

Low on travel time, we took Interstate 40 through the forested Arizona mountains and actually managed to pass a Volkswagen bus and a Honda moped while traveling upgrade.

As we droned along the big 4-lane highway for the first time on our trip, it occurred to me that the worst part of Interstate travel is its predictability. At 10 in the morning we decided we would stop for the night in Flagstaff, and eight hours and approximately 18 million crankshaft revolutions later we did. Right on time.

Running downhill and out of the mountains the next day, we descended toward the California border and into the summer desert heat. At the end of the day we stopped in Needles for dinner and debated what to do, stop for the night or push on.

Normally we would have stayed in Needles overnight and crossed California the next day. But driving a British car across the Mojave Desert in July is sort of like sneaking a relative out of East Germany—your favorite old aunt, the one who talks too loud with her hearing aid turned off and likes to hit policemen with her umbrella. You don't want to do it in broad daylight, with everyone watching.

We lowered the top, changed into shorts and sandals at a gas station, bought a quart of Gatorade and slipped into the Mojave under the cover of darkness. Almost darkness. It was actually a brilliantly starlit night, with a hot western breeze coming straight out of God's own electric hand dryer. The Sprite was running fine (although the tach needle was beginning to send wildly exaggerated reports) and I wondered if top-down night travel might not be the best way to cross the country.

Then, on a long upgrade near Ludlow, the engine began missing, badly.

"Short?" Barb asked.

"Probably," I said. She was learning fast.

I stopped to look the engine over with a flashlight, checking the usual suspects, but couldn't see anything wrong. The miss sounded like a coil problem, so we pressed on, chuffing and bucking to Barstow before we'd had enough and found a motel at three in the morning.

When we started the car after breakfast the next day it ran fine, so we accel-erated up the ramp and onto I-15. Every few minutes the car would miss (i.e., stop running), then kick back in, just to remind us the problem had not gone away. But it stayed running, more or less, all the way through Victorville, up Cajon Summit and down the San Bernardino Mountains into the Valley of Brown Air and Palm Trees. We pulled into our driveway in Costa Mesa at noon on Sunday.

Home.

Normally when I get home from a trip I go straight into the house and relax, fix a drink, read mail, etc, and I don't want to look at the car or think about it for a day or so. But this time I immediately unloaded the trunk, pulled out suitcases, toolboxes, spare parts and all the other junk we'd accumulated along the way. I wanted to unburden the car, the way you might lift heavy saddlebags from a faithful burro that had carried you across the desert. You couldn't leave it loaded, standing in the sun, while you sat in the house having a drink.

When the car was unpacked, I tried to drive it into the garage, but the engine wouldn't start. I got out and looked under the hood, where I found that the coil wire was burned almost completely through the insulation. It had been arcing internally, all the way from Ludlow. Maybe all the way across the country. I closed the hood and pushed the Bugeye that last 10 yards into our garage. Before going into the house I looked back at the car and said, "Close enough."

JEEPING BAJA

Part I, in which a large, nearby peninsula is discovered

BY PETER EGAN

THE FORGOTTEN PENINSULA

EL ROSARIO
EL MARMOL
Bahia de Los Angeles
PUNTA PRIETA
GUERRERO NEGRO
EL ARCO
SANTA ROSALIA
SAN IGNACIO
MULEGE
LA PURISIMA
COMONDUS
VILLA CONSTITUCION
MISSION SAN LUIS GONZAGA
LA PAZ

PHOTO BY JAMES T. CROW

PHOTO BY JAMES T. CROW

PHOTO BY J. T. CROW

PHOTO BY JAMES T. CROW

Dave Ekins and Bill Robertson, Jr set a 39-hour, 54-minute record to La Paz in 1962 on their Honda 250. Bruce Meyers and Ted Mangles lowered the record time to 34 hours, 45 minutes with Old Red in 1967, just before Spence Murray and Ralph Poole drove their Rambler American down in 31 hours flat. Even a TR3 joined the movement. Ken Kilfoyl and Floyd Fessler (#10) stop for fuel during the inaugural Mexican 1000 Rally in 1967.

Ruins of the old onyx quarry, El Marmol, closed 1959.

40

I 'VE ALWAYS WANTED an adventure to begin, as it does in so much Romantic fiction, in the walnut-paneled chart room on the second floor of an explorer's club, preferably in London. You've seen the room: overstuffed armchairs, leather ottomans brought back from the Dardanelles, Zulu spear on wall next to portraits of Livingstone, Burton and the Queen; Edward Whymper's ice axe in a glass case, an oversized slightly yellowed globe whose empty spaces bear simple names like The Congo, and a wall full of morocco-bound books written between bouts of malaria by the very men who pioneered sturdy footwear and first distinguished between the White and Blue Niles.

Unfortunately, I live in Orange County, California where there are a lot of transmission repair shops but very few dark-paneled explorer's clubs. Adventures in our neck of the woods are more commonly hatched over margaritas or Carta Blancas in the nearest Mexican restaurant with good air conditioning. Which is exactly the sort of place this one began, about one year ago.

I was having lunch at a local spot called El Ranchito, which has pretty good food, waitresses in Carmen Miranda outfits, sans fruit chapeaux, and plenty of brightly colored parrots painted on the stucco walls. Across the table from me was James T. Crow, whom readers will recognize as a former editor of R&T, as well as the founder and publisher of *Pickup, Van & 4-Wheel Drive* magazine. Jim is also the author of a book called *Baja Handbook for the Motorist in Lower California*. Added to these distinctions, he is a neighbor of ours and likes to eat Mexican food, so we get together for lunch fairly often.

That particular day we had been exchanging views on T.E. Lawrence, Edward Abbey and other desert writers over enchiladas, discussing at some length our mutual fascination with the drier portions of the earth's surface, which we agreed had something to do with growing up in humid midwestern summers. Toward the end of our conversation, Jim suddenly gave me a thoughtful look and said, "I think I've got a book you should read." At our next meeting he casually slipped me a copy of *The Forgotten Peninsula*, by Joseph Wood Krutch. A book about Baja.

He might as well have given young Tom Edison a book called *Careers in Electricity*. Krutch painted a vivid picture of this 800-mile-long peninsula: its strange desert and mountain landscapes, the many failed attempts to exploit the place, leaving abandoned mining towns, ranches and Jesuit missions stuck in all kinds of remote and inaccessible places, hidden valleys with lush date palm oases, caves with Indian petroglyphs and coyotes howling at the moon. Dangerous reading, in other words. It led not only to a burning desire to throw down the book and rush

off to Baja, but also to the discovery that there was a sizable body of literature on the place. Our local library had a whole shelf of Baja books.

It's been said that everyone who ever went to Baja has written a book about it, which is not quite true, but very few professional writers have resisted the temptation. John Steinbeck, Erle Stanley Gardner, Jack Smith, Walter Nordhoff and dozens of others had all fallen under the Baja spell, following in the footsteps of Hernando Cortés biographer, Bernal Diaz, as well as the Jesuit *fraile s* who founded the missions and turn-of-the-century outdoorsmen like Alfred North. There were also dozens of modern travel guides to the peninsula, the most detailed being Wheelock and Gulick's *Baja California Guidebook*.

After thumbing my way through that batch of eyestrain, I also discovered that the R&T library had a great collection of race reports on the Baja 1000. Between *Road & Track*, *Cycle World* and *Pickup, Van & 4-Wheel Drive*, nearly every aspect of these races had been reported.

Racing in Baja, I learned, began in the early Sixties with a few adventurous individuals making lonely record runs from Tijuana to La Paz. In 1962 Dave Ekins and Bill Robertson, Jr, riding Honda 250 scramblers, covered the 1000-mile route in 39 hours and 45 minutes. Then in 1967 Bruce Meyers and Ted Mangles drove Old Red, the original Meyers Manx dune buggy, to a new record of 34 hours and 45 minutes. Two months later Spence Murray and Ralph Poole blasted from Tijuana to La Paz in 31 hours flat, driving, of all things, a Rambler American; Melmac turquoise green, with a 199-cu-in. inline-6, like the one your grandma had.

These small outbreaks of Baja fever turned into a general epidemic, and in late 1967 Ed Pearlman and Don Francisco formed the National Off-Road Racing Association (NORRA) and ran a full-fledged, timed race called the Mexican 1000 Rally. The inaugural race was won by Vic Wilson and Ted Mangles, again in a Meyers Manx, in an elapsed time of just 27 hours and 38 minutes. In 2nd place, an hour and 10 minutes behind, was a Husqvarna 360 co-ridden by motorcyclists Malcolm Smith and J.N. Roberts. Sixty-eight vehicles started the race—buggies, motorcycles, Jeeps, Toyota Land Cruisers, a Ford Ranchero, a Saab 96, a Datsun Patrol and several Jeepsters. Thirty-one finished, leaving a wake of dust, shredded tires, bent wheels, broken axles and clogged air filters from the border to widely varied points south. A racing tradition began that has continued, in various shortened and looped versions, right up to the present.

Reading about it now, it seems to have been a terrific time, part fad and part obsession, when half the crazy car people in

Jeep on dunes, south of San Quintin.

southern California and elsewhere turned their attention away from road racing, go-karts or whatever else had lost its allure and found themselves held in the sway of this strange piece of land, twice the length of Florida, appended to the other corner of North America. Steve McQueen and James Garner raced there; Bruce Brown, coming off the success of his surf documentary film, *Endless Summer*, made a motorcycle movie called *On Any Sunday* with some beautiful footage of Baja.

Pretty much saturated by all this wonderful lore, I decided it was time to see Baja for myself; to stand on the bay where Cortés landed at Loreto, visit the abandoned Jesuit missions and drive across the strange landscape of the legendary Mud Flats where all those buggies, Jeeps and bikes had raised such storms of dust. I'd seen the movie, read the books and steeped myself in the old magazines, and now it was time to go.

I'd actually been to Baja three or four times on short motorcycle trips, weekend raids across the border into what Jim Crow calls "suburban Tijuana," but never deep into the peninsula where the Real Baja is said to be found. The main highway, Mexico 1, was now paved all the way from the U.S. border to the tip of Baja at Cabo San Lucas, blacktop running down the peninsula like a long spinal cord. Any car, I was told, could negotiate the main highway, but to reach the nerve ends—the remote beaches, the abandoned missions and ruined mines—I would need the right vehicle. An off-road device.

This was a problem, of course, because until recently my idea of off-road driving had been limited to locking up the brakes on my Formula Ford and skidding off Corner 7 at Riverside. Low ride height, in fact, was the keynote of my entire vehicle collection; I didn't own any 4-wheeled vehicle that wasn't spooked by the mere rumor of speed bumps and manhole covers.

Lou Bintz to the rescue. Lou is an off-road racer who happens to work for AMC/Jeep/Renault, and through his good graces I was able to wangle the loan of a Jeep from his fleet of test vehicles. He had a CJ-7, he told me, a bright red Renegade with a soft top, removable canvas doors, 5-speed transmission and the 258-cu-in. in-line-6. Not to mention a tape deck and air conditioning. I could pick it up from Brian Chuchua Jeep in Placentia, not far from our office.

Brian Chuchua, owner of the dealership, is something of a Baja legend himself, having sponsored and raced Jeeps down there for years, when not exploring the peninsula with one of his own helicopters. When he heard where we were going with the Jeep, he took it to a nearby shop and had a Rickard rear bumper installed. This is a heavy-duty affair that holds three jerry cans of gas and opens like a tool chest to carry an axe, shovel, spare oil cans or whatever else you want to put in it. With the bumper installed, I headed onto the freeway and drove home.

My experience with Jeeps, up to this point, was rather sporadic. My dad had a Jeep when I was a kid, until a wheel fell off and it overturned on the hill in front of our house; a mechanic had done brake work that day and had left off a rear axle nut. Later, I drove a Jeep all over Vietnam. This one worked fine until it was blown to smithereens by a mortar shell, happily while I was standing nearby rather than sitting in it. Neither of these problems was the fault of the Jeep, of course, and the second one wasn't a real Jeep anyway, but an M-38, made by Ford.

The red Renegade was by far the newest Jeep I'd ever been in, and it was a joy to drive, even on the freeway. The big six had tremendous torque and cruised easily in 5th gear, the top fluttered lightly in the wind; and the high seat offered a commanding view of the road. I was pleased to see that the Jeep had retained so much of its honesty and simplicity. With its big, simple instruments, plain steel dash and basic nuts-and-bolts chassis, it was the elemental utility machine, designed to do a single job well. There was a leftover American ethic lurking somewhere in the Jeep; I couldn't put my finger on it, but it had something to do with Glenn Miller and forest rangers.

With everything falling nicely into place, I decided it was time to line up a co-driver for the expedition, so I called my old friend Patrick Donnelly. Pat is an attorney who works for the Public Defender's office in Madison, Wisconsin. We grew up together in the same small Wisconsin town and were roommates in college. In the autumn of 1968 we took a 3000-mile motorcycle trip through Canada. In the rain. I figured after 17 years he'd probably forgotten and was ready for another trip.

"Just tell me when you want to go," Pat said, "and I'll clean up my trial calendar."

"How about two weeks at the end of April, before it gets too hot down there?"

"Perfect."

In getting ready to go, I prepared a long checklist, working mostly from the recom-

mendations in Jim Crow's *Baja Handbook*. Making numerous trips to the supermarket, sporting goods stores and a large off-road specialty shop—Dick Cepek's—I gathered together a large pile of supplies in the middle of our garage floor; camping gear, food, a Coleman cooler, water jugs, tire repair kits, maps, axe, shovel, a Hi-Lift jack to get out of sand, a tow strap in case we couldn't, flashlight, snake-bite kit, etc. Jim Crow said that two small lawn chairs and a folding table (which he lent me) were indispensable for a good Baja campsite, so I added those to the pile. Having camped out of a backpack rather than a vehicle most of my life, this seemed like wild luxury. With all the obvious stuff accumulated, I got together with several of the old Baja hands in our neighborhood for some last-minute advice.

"Water," Larry Crane said. "Fill up your canteens every chance you get. As long as you have enough water in Baja, all your other problems are secondary." Larry is the assistant art director at R&T, and a veteran of many trips into the peninsula with his Toyota Land Cruiser.

"Tires," Ted Mangles told me over lunch one day, "are your most important resource in Baja." Ted and his co-driver Vic Wilson, as I mentioned, won the first Baja 1000. "When you are out of tires in the middle of nowhere in Baja, you're out of luck. It's a big place."

"Just make sure you have plenty of gas," Jim Crow said. "There are gas stations scattered all over Baja, but you can't count on them having gas or having the electricity to pump it. The power goes off and on a lot in those little villages and no one is quite sure why."

"Please don't go," Pat Donnelly's mother said, just before he got on the plane for California. She was worried. Two Americans had just been shot dead in Guadalajara over some kind of drug misunderstanding, and there were ugly scenes at the border in Tijuana because an American border guard had shot a boy who was throwing stones at him. An American narcotics agent was also missing in Mexico and President Reagan said the Mexican government wasn't trying hard enough to find him.

There was no shortage of advice, pro and con, on the merits of travel in Baja. Advice from people who had been there many times was positive and upbeat, concerned mostly with maps, equipment and the proper provisions. Those who had never been south of the border, on the other hand, viewed the place with great suspicion, raising questions of bad water, banditos, unpredictable *Federales*, Mexican jails where they throw away the keys, scorpions, rattlesnakes under every other rock, cocaine smugglers and restaurants where the kitchen help did *not* wash their hands before returning to work. The old Baja hands chuckled over most of these stories and said any rumor that kept more tourists out of Baja was a good rumor. "Bad roads, good people," they said.

Good or bad, Donnelly and I were ready to go.

Pat flew into town on a Thursday night, and the next day we loaded the Jeep. Before leaving, we made a last run to the sporting-goods store and bought a couple of items labeled as "GI sleeping bag covers." We planned to sleep outside, under the stars, and wanted something to keep the mist off our sleeping bags for those nights when there were no stars.

When we unwrapped the sleeping bag covers at home, they turned out to be army-surplus body bags. One each. Olive drab. Donnelly and I looked at each other and shuddered. They were an ugly symbol from an earlier part of our lives.

"I'm not sleeping in any goddamn body bag," I said.

"Maybe if it rains we could just put them over us," Pat offered, "or not zip them all the way up . . . "

In the end, we refused to be intimidated by cheap theatrical omens and took the bags along.

I said goodbye to my wife, Barbara, and we left on a Saturday morning in April. Driving south through San Diego, we stopped on the American side of the border to fill up with gas, change money (242 pesos to the dollar that day) and buy Mexican car insurance. The law requires that you have liability insurance when traveling in Mexico, and you can buy it at any number of agencies at the border for about $7.00 a day. We were waved through customs and cruised around the tacky outskirts of Tijuana and followed the coast down 60 miles of mostly divided tollroad to Ensenada.

Ensenada is a large town (or small city) on the coast, far enough from the border to avoid the gawk-and-run crowd Tijuana attracts, but close enough to be a weekend tourist town for Americans, as well as Mexicans. It's a lively place, and I always like being there. We had a happy hour of free margaritas, courtesy of our hotel, then went out for what was probably a perfectly nice dinner at some sort of restaurant. Later we wandered over to Hussong's Canti-

na for the obligatory ritual beer. Hussong's is a hole-in-the-wall bar, widely celebrated on bumperstickers, that has the good fortune to be popular because it's always packed and vice versa. It's the kind of bar where you can pass out standing up and not fall down because there's no room. And a good thing, too, because you wouldn't want to see what's on the floor.

We awoke to a gray, misting 2-Excedrin Sunday morning and drove out early through the nearly empty streets of Ensenada. We passed the newly planted vineyards near San Vincente and climbed out of a broad valley into mist-shrouded mountains. Green with spring rain, they looked like a cross between Wyoming and the Scottish Highlands, with a little Peru mixed in. We drove down Mex 1 for an hour, then passed a road sign for the little village of San Telmo. We were now farther south than I'd ever been before. The road ahead was new territory, all unknown.

"This is it," I said to Pat.

He downshifted for a tight corner, a *curva peligrosa*, and looked across at me. "This is what?"

"The Real Baja."

By mid-morning the road dropped back to the coast, through the born-on-the-highway clutter of San Quintin and then past miles of beautiful sand dunes, into the town of El Rosario.

El Rosario is a little town where the paved road used to end, an old jumping-off point for the hard-core 4wd traveler, and it is still a signpost for the demarcation line between the relatively populated north and the great empty desert of central Baja. We fueled the Jeep and stopped at a small grocery store for ice, which I learned from my Spanish phrase book (for idiots who should have paid attention in high school Spanish but didn't) is called *hielo* and pronounced *yay*-low and is a most important word to know in Baja. We also bought some Mexican beer and bottles of those two related distillates, tequila and mezcal, for later comparison purposes.

On the way out of El Rosario we stopped for lunch at a small roadside cafe with some dogs on the doorstep. The place had mariachi music blaring from a radio in

Campsite by well, El Marmol.

the kitchen. There was one other customer, a small, unshaven man who sat in the next booth. As we ate, the man turned completely around in his seat and stared at me over Pat's right shoulder, which gave me the unnerving sensation of talking to a 2-headed person. I stared the guy down maybe four times, but the effect never lasted long and the face reappeared over Pat's shoulder. I told Pat it must be nice, having a talking head and a listening head. I knew which one was the real Donnelly, though, because the guy was puffing on a cigarette and Pat doesn't smoke.

South of El Rosario the road turned inland and the country suddenly became true desert; mountain and arroyo country of bleached gravel and stone. Giant cardón cactus appeared—the arms-raised variety cartoonists use to depict the West—and just a few miles out of El Rosario we saw our first boojum. The boojum, or *cirio* (candle), as the Mexicans call it, is an odd, carrot-shape plant that has small, spiny branches and can grow to be 40 or 50 feet tall. The tree has become the symbol of Baja because it grows naturally nowhere else. Even in Baja it's limited to a radius of about 125 miles. Boojums curl, twist and branch off into all sorts of bizarre shapes and look like John Wyndham's Triffids, or creatures from another planet; ambulatory beings that have stopped moving just long enough to watch you drive by.

Late in the afternoon, 250 miles south of the border, we turned off the main highway for the first time, heading east into the hills. We wanted to camp at a place Jim Crow had told us about, an abandoned onyx quarry called El Marmol.

The dirt road didn't really require 4-wheel drive, but we got out and engaged the front hubs anyway. The road to El Marmol was 18 miles of loose sand and gravel, over gently rolling terrain. We arrived at the old quarry just before sunset.

Huge slabs of yellow and white onyx were scattered around, mixed with the bleached-out bodies of old American sedans, mostly from the Thirties. The walls of an old schoolhouse sat in the middle of a clearing, and nearby were the ruins of the quarry manager's house, which Jim Crow told me once had a shaded porch with a porch swing and whitewashed shutters. Now there was nothing left but a set of melting adobe walls and a crumbling chimney; the kind of place a Confederate soldier might have come home to after Sherman's March. Not far away was a cemetery of about 40 graves, some mere piles of stone with crude, tilting crosses, and others concrete crypts with glassed-in shrines for flowers and icons.

According to our Baja guidebook, the quarry opened in 1900 and closed in 1958, when the world discovered that plastic pen and pencil sets looked almost as good as onyx ones. They were cheaper, too, because no one ever died of heat exhaustion dragging huge slabs of plastic out of the middle of Baja.

We camped in an idyllic spot between an arroyo and the ruins of the manager's house. Almost idyllic. About 10 feet from our cots was an uncovered well that could easily have swallowed our Jeep without a trace. We couldn't see the bottom, but water was exactly 2.5 seconds, by faint pebble splash, away from the surface. "How far does a pebble drop in 2.5 seconds?" I asked Pat.

"I don't know."

"Where's Newton when you need him?"

"Dead."

We made a mutual pledge not to walk near the well at night.

After a terrific dinner of canned ravioli and chocolate pudding, Donnelly scrounged up a collection of mesquite and old chunks of timber for a campfire. With the first stars coming out in a brilliantly clear sky, we sat down in our lawn chairs to talk and to compare Gusano Rojo Mezcal, distilled in Oaxca, with Cuervo Especial Tequila, *hecho in la Rojena*. It was getting cold out, but the heat from the fire felt good. And smelled good.

Burning mesquite gives off a wonderful aroma that is closer to being a flavor than a smell; a sharp incense in the crisp desert air, startling in its contrast, like drinking cold spring water and discovering the taste of a brown, spicy Rheingau in the stream. We watched early evening satellites cross under a sky full of stars undimmed by smog and city lights, and toasted Jim Crow for insisting we take lawn chairs. Later, when the moon came up, we toasted him again, for telling us about this place. ⊛

JEEPING BAJA

Part II, in search of the Lost City of La Paz

BY PETER EGAN

W AKING UP IN the desert is a little different from waking up anywhere else. When the sun appears over a ridge on a cool morning, warmth immediately flows into everything, like current through a soldering iron or coffee conducting its heat into the handle of a tin cup. It's a kind of radiation so unimpeded by cloud or haze that it seems to beam straight through space and bounce off the core of whatever it hits, warming from the inside outward. There's none of that gradual, stingy morning warmth the sun doles out in more northern latitudes. After a cold night, the Jeep is warm to the touch in only five minutes; the rocks are warm, the Coleman stove and the coffee pot are warm, and the average sleeping bag is too warm to sleep in.

There's no sleeping after the first minute of sunrise, anyway. The landscape goes from cold and shadow to light and warmth with a bang, the sun coming up like something from a toaster. An English muffin, maybe. Or in this case, a Mexican muffin, if there is such a thing. So far we've seen only tortillas. I look over at the other cot, across the smoldering gray ruins of the campfire and see that Donnelly, too, is wide awake. I'm relieved to find both of us still present and accounted for, because we are camped next to a large black hole in the ground that used to be a well.

Donnelly looks over at me and says, "Any sidewinders under my cot?"

"Not today," I say, knowing he's only half kidding. He doesn't like snakes. He doesn't like scorpions, either. Neither do I, so we both shake our boots and peer into them before putting them on. Time to get up and hit the road. We tie our boots and start stumbling around to light the propane stove and make coffee.

Jeep? Desert? Sidewinders? Where are we, anyway? Start making sense.

According to our AAA map, we are camped at an abandoned onyx quarry called El Marmol, 2300 ft above sea level, some 260 miles south of the U.S. border. My old friend Pat Donnelly and I are on the second day of a 14-day trip through the back country of Baja California, that 800-mile peninsula of Mexican desert and mountains attached to the U.S. just south of San Diego.

The map of Baja, like the peninsula itself, is long and narrow; it's the kind of map you spread out on the ground rather than on a desk or table when you want to

show people where you're going, a sweeping gesture that always makes you feel like a Moroccan rug merchant showing off his wares. If we follow the yellow Magic Marker lines on our map, we have about 2400 miles to go, most of it on trails of dirt and rock. Parked nearby is our accomplice in this trek, a red 1985 Jeep Renegade borrowed from Renault/AMC/Jeep; 258-cu-in. six, 5-speed, soft top, removable doors. I'm not crazy about the big Renegade decal and the styling stripes on the hood, but surrounded by boojums and cactus on a sunny desert morning, the Jeep looks absolutely right; a rugged, sensible piece of camping equipment, rather than a thinly disguised family car or station wagon. It belongs in Baja the way Land Rovers belong in Africa.

Pat and I load our gear into the back of the Jeep, cover everything with a tarp to keep out the dust, lift the doors off and stack them behind the seats, and we are

on our way, off the road again. The Jeep whines up a steep embankment out of the arroyo, around the stone ruins of roofless buildings and past the cemetery, where concrete shrines and piles of stone mark the graves of about 40 people who bumped and jostled up miles of bad road to quarry onyx here and never jostled back out. Baja, we are learning, is littered with the failed efforts of padres, farmers, miners, developers and would-be industrialists. The peninsula, by its very harshness, has won a lot of these small victories against human settlement, which is part of the reason Donnelly and I are here. There's something in all of us that likes to run the risk of being taught a lesson.

The other reason we're here is that great prerogative of adult life called making up for lost time. If you struggled through the first half of this saga last month, you'll remember that Pat Donnelly and I grew up and went to college together in the midwest. This was a great place to live, but it wasn't exactly the epicenter of off-road adventure during the Sixties and early Seventies. We both missed out on what might be called the Golden Age of Messing Around in Baja; the early days of the Baja 1000 race and the flowering of off-road exploration in the peninsula as cult and avocation. Some of the early excitement and innocence of the movement had died down in the meantime, of course, but Baja was still there, waiting to be explored. So we planned a route, borrowed a Jeep, and Pat flew out from Wisconsin to join me on the trip. It was the second night of our drive that landed us, off the road, at El Marmol. Camped between a bottomless well and a cemetery.

Leaving El Marmol that morning, we drove 18 miles by dirt road back to the main highway, Mexico 1. It was a sunny, warm April morning, not long after the last spring rains, and the red flowers of the Indian paintbrush were in full bloom along the road.

Right after hitting the blacktop of the main road, we stopped for breakfast at a little rancho called Tres Enriques; the Three Henrys. There were no Henrys in sight, but a nice woman served us a good breakfast of huevos rancheros, tortillas, frijoles and coffee for 1200 pesos, or about $4.94 for both of us. Some young Americans pulled up in a Blazer and shuffled in, groaning, as though stiff and sore from sleeping in some uncomfortable place af-

EL ROSARIO

EL MARMOL

Bahía de Los Ángeles

PUNTA PRIETA

GUERRERO NEGRO

EL ARCO

SANTA ROSALÍA

SAN IGNACIO

MULEGÉ

LA PURÍSIMA

COMONDÚ

VILLA CONSTITUCIÓN

MISSION SAN LUIS GONZAGA

LA PAZ

ter drinking too much. One was giving the others a sort of Cliffs Notes/Classic Comics lecture on the geology of Baja. "You can find low-grade gold ore in the rock around here, but it costs too much to get out," he told them, holding up a small chunk of granite.

"How do you know so much about gold all of a sudden?" one of them asked.

"Last night he knew all about snakes," the third one mumbled.

"Rocks are the one thing I *do* know about," the geology guy protested, slightly hurt. We left them to work this one out, heading south toward Catavina, into a landscape suddenly populated with white-barked elephant trees.

South of Catavina we made a detour to cross Laguna Chapala, a large dry lake, and drove across its flatness toward a permanent mirage of water. The Indians, legend has it, saw mirages of Spanish galleons floating on these dry lakes long before they met the Spaniards. If it's true, it must have been a chilling vision. The south end of Laguna Chapala is supposed to be a good site for finding arrowheads and other projectile points. I got out with my rock hammer and looked around while Pat drank a beer, but found only a bleached white bone that I positively identified as a femur from either a saber-toothed tiger or a dairy cow.

We stopped at the junction for our turn-off toward the coast and filled the Jeeps at a Pemex station, one of those rustic outposts of the state-owned petroleum industry. These stations usually have two pumps, one for the low-octane *extra* and the other for the even lower octane *nova*. Mexican gas is famous for its ability to cause detonation in any engine with a compression ratio greater than 1:1. This station had only *nova*, in a pump powered by a small Honda generator with a blown muffler. "I didn't know Honda ever made a piece of equipment this noisy," Donnelly shouted at me. We got 35 liters for 1950 pesos, or about 87¢/gal. The Jeep was getting 18 mpg, combined city/highway/boondocks, non-EPA cycle.

Twenty-five miles south we pulled off the highway, climbed out of the Jeep to engage our front hubs, put it in low-range 4-wheel drive and began a 20-mile climb into the mountains, toward the abandoned Jesuit mission of San Borja. This was the first really rough road we had faced with the Jeep—a dirt trail of sand, rain ruts, rocks and steep climbs out of stream beds—and it was over this terrain that the Jeep really came into its own.

You sit up high in the Jeep, whether driving or riding shotgun, in vertical seats, like men in a horseless stagecoach. Not quite horseless; there's a hood out there hiding the team from view. You can't see them sweat or hear them wheeze on a long, sandy uphill run, so you watch the temperature gauge and listen for the hot

rattle of Mexican gas in the cylinders. But the temperature gauge stays on Normal and there's no protest from the engine.

The Jeep's 258-cu-in. six is the kind of engine that used to make Americans like their cars; loads of bottom-end grunt, yet smooth when revved, equally willing to walk or run. There's no waiting around for turbos or the miracle of torque multiplication through smaller pistons and lower gearing. The power is right there, on tap as needed, ready to spit rocks, chug along through sand or just idle. It's an engine a machinist mate in the Navy would love. After a long grinding run up a hillside, you want to do something nice for the big six—oil its rockers, tighten the pillow blocks; something—but there's nothing to do. The engine is just fine.

The Jeep has a selector for both high and low range, of course, and we used low range on tight, steep mountain trails like this one, mainly because it gave us a tightened-up set of ratios for slightly different traction problems; low gear for heavy sand or crawling over oil-pan threatening rocks, 3rd for rutted hairpins and 5th for going up to 40 mph on straight sections. In low range it's possible to drive so slowly that you begin to wonder if you are actually traveling or just operating a geared-down traction device that pushes solid ground away from its tires, causing the earth to rotate beneath them. The gears whine like some kind of celestial clockworks, and you have the feeling that by shifting down a gear, you could delay nightfall for the entire planet.

After about two hours of very slow climbing, the Jeep launched us out of the valley and into the end of a high box canyon. There, surrounded by hillsides of boojum trees and giant cardón cactus, was the Mission of San Borja, an imposing Moorish-style church of white stone. Smack dab in the middle, as we say, of nowhere. The mission was started by the Jesuits in 1758, but Dominicans built the present church in 1801. There were 3000 Indians in the area when the Jesuits arrived. Sixty years later, there were 175 Indians, most having died of European diseases. The mission was abandoned in 1818. This pattern was repeated at about 30 other missions in Baja, eight of which are still standing, in various states of restoration. Others are completely gone, or have left only traces of adobe walls.

Pat and I explored the mission and its grounds, noticing suddenly that we were doing so under the watchful eye of a man who emerged from a little ranch house on a nearby hillside. I walked up the hill, shook hands with the man, and then dazzled him with my high school Spanish by saying, loosely translated, "Nice mission." He agreed that it was nice, and we smoked a couple of my cigarettes before I said *adiós*.

Leaving at sunset, we thrashed our way

back down the mountainside in the dark, with bats and owls fluttering across the edges of our headlight beams. We got into the little gulf coast village of Bahia de los Angeles at 9:00 p.m. and checked into a slightly dismal motel, the kind originally built to ruin the honeymoons of luckless couples who hadn't planned ahead. We took cold showers from a nozzle that was more like a slow leak in the plumbing than a deliberate attempt to spray water, then climbed into what a friend calls "taco beds," because they fold in the middle.

"Seems like we've been here before," Pat said.

"Yeah," I said. "Montreal."

In 1968 Pat and I took a 3000-mile motorcycle trip through Canada together. In Montreal, through luck and frugality, we managed to find the worst hotel in town, a dark place filled with hookers and alcoholics. And 18-year-old motorcyclists with no money.

In the morning we got ice, fuel and had breakfast on the porch of a nice cafe overlooking the bay, then spent the morning driving a gravel road down the coast toward a seaside resort called Punta San Francisquito. On the way in, we came across a broken-down truck. The owner, a big, gregarious man who introduced himself as Jorge, had a hole in his radiator and was trying to patch it with an old screw and a piece of inner tube. I got some epoxy resin and hardener out of our tool kit, mixed up a batch on a flat rock, and we smeared it over the hole in his radiator. I asked in my makeshift Spanish if he had enough water to fill the radiator. "*Sí,*" he said, laughing. He opened up the back of the truck and showed us his cargo. It was a load of ice. "*Tengo mucho.*"

We all had a beer, then Pat and I shook hands with Jorge and left. When we drove away, the epoxy still hadn't hardened. Pat said, "What do you bet, as soon as we're out of sight he scrapes that miracle space-age crap off his radiator and puts the piece of inner tube and the screw back on."

"He might have to," I said. "I can't remember if that's 5-minute epoxy or 24-hour epoxy."

Punta San Francisquito is a rustic little row of thatched huts gathered around a lovely sand cove. The place was built mostly for fly-in traffic and has an airstrip that allows you to taxi your plane almost up to the backdoor of the cabins. There's also a semi-outdoor bar and a small cafe attached. We had turtle soup for lunch, then relaxed for the rest of the afternoon. Pat stretched out in the sun on the beach while I, already slightly sunburned, sat on the shaded front porch of our hut, reading a book and working on my autobiographical screenplay, *Albino Beach Party*.

In the evening we cooked up a delicious batch of Dinty Moore beef stew and drank a bottle of Santo Tomas Vino Tinto, a wine we had bought in the village of Santo

Above: Bougainvillea at El Acro, mission interior at San Borja, and author identifies fossil bone as being from dead animal. Beach at Punta San Francisquito (right), and giant cardón near Cataviña.

Tomas, near Ensenada. We proclaimed it a wine that not only traveled well, but jounced, lurched, tipped, careened, bounced, bottomed out, shook and shuddered well. Amazing stuff; the perfect Baja Jeep wine.

Leaving early the next morning, Pat and I agreed that we were getting to be landscape junkies, not entirely happy unless we were rolling, doors off, into new territory. An afternoon and evening at even an idyllic spot like San Francisquito strained our tolerance for staying planted. It felt good to climb into the Jeep and go every morning. After three days, it was beginning to feel like home.

Right before leaving on the trip we had installed a dash-mounted compass, almost as an afterthought. On this stretch of road, and for many days ahead, it proved to be invaluable. We had three different

road maps of Baja, including a route guide to the original 1967 Baja 1000, plus an aviation chart of the peninsula. The AAA map was by far the most useful, but even that did not begin to show the amazing proliferation of roads and trails around Baja. It would indicate a single road winding between two remote villages, but in driving the route you ran into an endless succession of forks in the road, 3-way intersections and dead ends. Without a compass, good hunches, plenty of gas and a willingness to admit you were wrong and backtrack down 10 or 15 miles of bad road, you could get very lost in parts of Baja. And very thirsty.

We navigated, Columbus-style, on a sea of open desert all morning until the little town of El Arco appeared on the horizon. We pulled into the town square and stopped at a little, shaded general store for provisions (*cerveza y hielo*).

El Arco used to be a checkpoint for the Baja 1000, and the cash register at the store was plastered with old racing-team stickers. Without cars and bikes blasting through town, the stickers seemed oddly out of place, the only clue that El Arco was not, in fact, a movie set designed by Sergio Leone for Clint Eastwood. We walked out onto the town square at noon, and there was no sign of life except a dog crossing the dirt street and an old woman sweeping in the shaded vestibule of the church. The adobe houses had that mute,

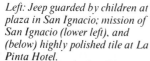

Left: Jeep guarded by children at plaza in San Ignacio; mission of San Ignacio (lower left), and (below) highly polished tile at La Pinta Hotel.

shuttered look you find in the villages of rural Spain, where you aren't sure that anyone lives there. When we left town, I had the feeling we were being watched from dark doorways.

We got lost only once driving cross-country toward San Ignacio, and then only for a couple of hours. The road we were on tapered off into a goat path, minus the goats and, eventually, minus the path itself. A road to somewhere people no longer wanted to go. An abandoned ranch or mine, probably. We backtracked past the ruins of a ranch house and corral, drove due west by compass for an hour and hit the main highway again late in the afternoon.

Toward evening, we drove into San Ignacio, one of the prettiest towns in Baja, an oasis of green, surrounded by barren desert. It's built down in a well watered arroyo, and the road into town is almost a jungle of date palms, vineyards and fruit trees. There's a shaded plaza in the center of town, dominated by one of the best-preserved mission churches in Baja. Pat and I sat on a parkbench under a giant tree to admire the church, while three young boys on bicycles shyly inspected our Jeep and argued quietly over the function of the locking mechanism that held our jerrycans in place. After living in our neighborhood in southern California, the manners of Mexican children are a delight to behold. They have not been taught to believe they are small centers of the universe.

It seems to be a contagious condition, caught from their parents. The Mexican adults we met in Baja tended to be polite, dignified, honest to a fault and helpful all beyond the call of duty. People you meet on the streets in the small villages are sometimes slightly guarded and shy until

you smile and say, "*Buenos dias.*" Once you have proven yourself human by the simple act of being friendly, the shyness disappears and your goodwill is reciprocated with a cheerful greeting. There are probably some unpleasant people living somewhere in Baja. We just never succeeded in finding any.

We found a room for the night at La Pinta Hotel, a very nice place with a Spanish-style tiled colonnade around a pool. There is a chain of these hotels in Baja, built to handle the increased American tourist trade when the Transpeninsular Highway (Mex 1) was completed. They don't have the rustic, one-off appeal of Baja's older hostelries, but when you've been driving a Jeep through the dust and camping for a couple of days, a working shower nozzle and hot water are not all bad.

Early in the morning we drove out onto a high plain with the volcanic peaks of Las Tres Virgens rising above the clouds, Kilimanjaro-like. We pulled off to eat breakfast at a little place called Rancho El Mezquital, a tilting building with a log frame and palm-thatch walls and roof, held up by support ropes to the remains of an old Ford station wagon. It looked like the sort of place where Mad Max might have had breakfast in *The Road Warrior*. An aged but handsome woman ducked through the low doorway to the kitchen to see what we wanted to eat. At most ranchos in Baja there is no menu. The only question is whether you want to eat. If you do, there may be a couple of choices. At this place the choice was huevos rancheros or tortillas con machaca. Bacon and ham were also available, she said. We ordered the huevos rancheros, which, when they came, we voted Best of Trip.

While we ate, some hens and several chicks came out of the kitchen, pecked around our table, then left by the front door, which had a blanket tacked across it. Then two cats came by casually, one at a time. Then a dog appeared, stopping by our table to scratch fleas. The chickens came through again, followed some minutes later by the cats. The dog left by the front door a second time, then reappeared from the kitchen. It was obvious the animals were doing laps. When we left, the chickens were ahead by nearly a full lap.

Near the gulf coast we descended into Santa Rosalia, an old mining town built by the French and full of 19th-century company-town equipment and buildings, including row houses and a galvanized iron church; a little of old New Orleans mixed with Key West and Pittsburgh. We walked around town for a while and bought ice, ice cream and some cookies that looked like dog biscuits, or worse, but tasted great. Baja is rich in *panaderias*, bakeries that make some of the best bread and pastries I've ever had. More than once on this trip, we loaded up on cookies, bread, or delicious flat things we called elephant

City limits, San Jose de Comondu.

Campsite on beach, near Cabo San Lucas.

ears and snacked out of a white paper bag rather than stop for lunch.

We drove down the coast for 30 miles, stopping in Mulege for a beer. This little gulf town seemed to have a fair number of hip young Americans shopping in the stores or just hanging out; the same crowd that was in Marrakesh 15 years ago. I always wondered where they'd gone. Farther down the coast we drove along the shores of Bahia Concepción, one of the loveliest bays in Baja. Unfortunately, we were not the first to notice, so there were campers and tents cheek-by-jowl on every sand beach. Following Jim Crow's dictum that any Baja campsite where you can hear or see a car, truck or other human being is a bad campsite, we drove 30 miles off-road to the tip of the peninsula that forms the bay. We set up camp near the beach, at the ruins of an old manganese mine the U.S. Army built during World War II.

That night I cooked up a truly wonderful dinner over our propane stove. We'd had a large can of Spam in our food box and Donnelly, being no great fan of Spam (having overdosed as a child, apparently), had demurred each time I suggested having it. I finally won him over, however, by promising to turn my considerable talents as a gourmet campfire cook on this can of meat. I came up with an impromptu recipe we christened "Spam Pedro." It was so successful that Pat had two helpings, and I've decided to pass it along.

Spam Pedro

2 white Mexican onions, chopped
1 7-oz. can of Spam
1 12-oz. can Green Giant Mexicorn with pimento
1 4-oz. can Clemente Jacques Salsa Estilo Casero
4 shakes Buffalo brand salsa picante, available in El Arco general store

Fry onions and diced Spam in 3 tablespoons of oil until onions are tender and Spam gets crisp around the edges, then add other ingredients. Salt to taste.

Serve with beer and tequila.

Recommended dessert: 5-oz. cans Del Monte Chocolate Fudge Pudding.

Sleep outdoors.

After dark it got very chilly out, so we put on wool shirts and sat close to the fire. Campfires of desert wood burn with an in-

tensity that seems to go beyond their dryness, returning all that stored sunlight to the cold night air all at once. The flame doesn't warm the air, but throws radiant heat 6 or 10 ft out from the center of the fire, warming your face and one side of your tin cup of tequila, while the back side of everything remains as cold as the dark side of the moon. It's space heat, delivered from space, and in the desert at night you can easily imagine it shooting back into space, dissipated and lost without ever having warmed the earth at all. Energy on a short visit from a nearby star. I suspected there might be a secret of physics locked up in the relationship between campfires, the full moon and the desert sun. Was the universe nothing but a giant billiard table for light and energy?

My search for this elusive truth was interrupted when I got up for another cup of tequila and discovered the toe of my left boot was glowing in the dark like a red coal, from being too close to the campfire. Furthermore, part of the clear rubber sole had melted into a small pool, picking up twigs and chunks of foreign material so that it looked like a small mat of plastic party vomit. I took out my Swiss Army Knife and trimmed it off, thinking before I went to bed that you have to admire an army with a corkscrew on its official knife. When I was in the army, the only knife they gave me was a bayonet.

In the morning my eyes popped open exactly at sunrise over the gulf. I lay in the bag for a while and watched a turkey buzzard hang on the wind, right over our camp, maybe 50 ft up. He was looking us over, balancing himself with small adjustments of the wings and tailfeathers. Donnelly and I had both covered our sleeping bags with army surplus body bags, which had been sold to us as "GI Sleeping Bag Covers." I moved my arms out of the bag, and the buzzard glided forward over the ridge behind us, having lost interest. Donnelly was watching, too.

"Good to be alive," I said. We made boiled coffee, then spent the rest of the morning driving out of the peninsula, past small, makeshift fishermen's shacks on the coast. On the way out we saw a coyote,

looking very healthy and loping through the brush, and an unusually large number of buzzards and road runners.

Afternoon found us on what was probably the most beautiful road on the trip, a steep climb from Loreto, on the coast, to the mountain mission town of San Javier. The road clung to the sides of a chasm-like valley with date palm groves far below, fed by a brook that tumbled into clear, deep pools and waterfalls; the kind of road from which ivory hunters and gun bearers used to regularly plummet while looking for the Elephant's Graveyard. At the head of the canyon was a mountain peak of such geometric proportion that it looked almost like a man-made pyramid, a peak you would choose to be a Holy Mountain if you were inclined to look for mystery in nature, rather than just name things in Latin. We left the Jeep at one point and hiked up the stream bed to examine some small caves, finding what looked like genuine Indian cave paintings. These were mixed with paintwork of more suspect origin, like one inscription that said "Ramone y Maria." Late Rustoleum.

As we came over a mountain ridge, the mission of San Javier loomed ahead, probably the best-preserved Jesuit church in Baja, and still used for regular church services. We sat on pews for a while in the cool, quiet church, taking it all in. Looking at these remote missions always brought the same obvious questions to mind: How did the Jesuits ever find this valley, so far up in the mountains? How did they ever persuade a population of hunters and gatherers to do all this work? How did they quarry all this stone and shape it so beautifully with limited tools in the middle of nowhere in 1744? And, last, where did their indefatigable energy come from? No matter what your feelings about religion and progress, the missions in Baja were a real monument to the power of human will, carefully focused. Contemporary accomplishments seemed feeble by comparison. The Jesuits made restoring a Jaguar look like something you could do on lunch break.

On the other side of San Javier, the road wound downward toward the Pacific coast

Approach to San Javier (above) and mission interior (above right); church is still in use. Jeep (right) sacrificing its third and last flat of the day.

through green, well watered cattle-ranch country. We had planned to camp somewhere in the mountains, but I have an aversion to camping in the company of cattle and cow dung. It reminds me of boyhood camping in midwest pastures when I really wanted to be on a Canadian lake or in the Rocky Mountains. Nothing quite so completely takes the romance out of camping as a bunch of cows standing nearby, chewing their cud and staring liquidly into your campsite. We ended up, five or six off-road hours later, spending the night in a cheap but presentable hotel in a place called Ciudad Constitución.

The area around Ciudad Constitución is the dullest and probably the richest part of Baja, agriculturally; a flat coastal plain full of wheat, agri-chemical smell, powerlines and cow towns full of pickup trucks. We put the Jeep in 5th gear, 2-wheel drive, and sped through this country as fast as possible, back over the mountains and toward La Paz.

Terminus of the old Baja 1000, La Paz is a good city in which to end up after a hard drive. It's a large resort town built on a peninsula so that the city looks back at the mountains of mainland Baja. The downtown hotels were full, but we found a room at La Gran Hotel La Paz, which was not a bad place for a poured-concrete, high-rise modern hotel. They had free drinks on a veranda overlooking the bay, a bossa nova band and many prosperous-looking Mexican guests. Pat and I drove into town and spent most of Sunday morning doing our laundry at a coin-operated laundromat. You could almost hear

the tiny screams when our socks and underwear hit the hot soapy water.

Resplendent in clean clothes, we went out on the town that evening, stopping by a few bars and dining on the terrace of an outdoor cafe in the park. It was a balmy, tropical night. We were just having coffee when a single voice rose above all the others in the cafe. We turned to see an American woman in bangle earrings and a large hat. She was asking the waiter, very loudly, if the restaurant's drinking water was bottled and if the fruits and vegetables were washed from bottled water or tap water, and if the ice cubes were made from bottled water. "*Sí, señora,*" said the waiter to all her questions. I suspect he kept his fingers crossed, or at least made the sign of the cross when he got back to the kitchen. We had been told the water in Baja was drinkable everywhere and hadn't found otherwise. The woman, however, had read all the travel books on exotic lands; she was nobody's fool and wanted everyone in the restaurant to know.

"A seasoned world traveler," Pat said dryly.

"Makes me wish we were back in the Jeep."

We were back in the Jeep soon enough.

We'd spent one week getting to La Paz (short of the old Baja 1000 record by about six days) and would spend another week getting back out. We would circle the very tip of Baja, drive up the Pacific coast, get lost in the plains north of Todos Santos, camp on beaches and on the surreal, endless mudflats of the central coast, have three flat tires in one day, run out of

spare tubes, backtrack out of the wilderness on three good tires to hitchhike 40 miles in search of an inner tube or an air compressor to seat a tubeless tire, find one in the third town we searched, seat the tire, hitchhike back with a large Mexican family in a Chevy van holding children on our laps, drive all night to Ensenada and generally have enough adventures to fill the average 7-minute Bob Dylan song. Among other things.

We got home on a Sunday afternoon, pulling into the driveway with the bright red paint of the Jeep almost obscured by layers of dust. In 2600 miles the Jeep had required two quarts of oil. Mechanical problems? I broke a snap on the convertible top, and after one particularly bad tank of gas at the end of the trip, the engine idled roughly and pinged. Otherwise, nothing. We never laid a hand on the Jeep, except to change tires. It actually came out of Baja as free from squeaks and rattles as it went in (more than you can say for my reading glasses and zoom lens, which shook themselves to pieces in the padded center of my camera bag).

We spent a quiet Sunday at home, resting and washing the Jeep, and then I drove Pat to the airport on Monday morning, so he could fly home. We shook hands and he said, "We'll have to do this again."

If he'd said that after our cold, rainy trip through Canada in 1968, we both would have laughed. This time it was no joke. Sooner or later, we'd be going back.

Baja was not an easy or luxurious place in which to travel, outside of the few large tourist towns. It was a country of fine margins; of roads that were almost not needed, plants that had just enough water, ranches of thin cattle that couldn't afford to lose one more pound, restaurants where a couple of customers—more or less—in one week could make all the difference between staying open or blowing away in the dust, and old missions saved by government money that came as rarely as desert rain but just in time to keep the walls from collapsing. In the central, arid country—where the peninsula was at its most stark and beautiful—life, buildings and towns were right on the edge. To travel there, you had to plan ahead and pay attention, like a sailor on the ocean or a pilot flying over new territory.

But knowing what was in Baja was like opening a wing in your house where the door had always been bolted shut and suddenly discovering a simpler and more charming set of rooms, left just the way they were in an earlier century; quiet, faded and a little dusty. There were people who didn't care for the furniture, of course, like the lady in the cafe. And that, possibly, was the best part. Visitors who didn't understand the place would never go back. In some respects, Baja was just like our Jeep; perfect in its own special way, but not for everybody. ◉

THE ROAD TO EVEREST

New Delhi to Kathmandu, via Oxford

PART I

BY PETER EGAN

PHOTOS BY THE AUTHOR

"**Y**OU AREN'T REQUIRED by law to have any shots to go to India and Nepal," the nurse told me over the phone, "but the Department of Health strongly recommends that you get immunized against typhoid, tetanus, polio, malaria, meningitis, hepatitis and cholera."

"Supposing I skipped the shots," I suggested. "Could any of those diseases be considered serious?"

There was a long silence on the other end of the phone.

"Only kidding," I said.

I got the shots, needless to say, and so did Chris Beebe, who was going with me, even though he'd never had a shot before in his life and had to have his hand patted at some length between each injection by a nice county health nurse in far away Wisconsin. Chris and I would be meeting

TO EVEREST

Our friends, Kirit Singh (left) and Param-Git Singh, chat on a stop near Moradabad as Chris Beebe prepares for his first-ever stint behind the wheel of a Hindustan Ambassador.

at O'Hare in Chicago and flying to New Delhi together.

Why?

Editor John Dinkel asked me that very question.

We were going to India because our mutual friend, Kirit Singh (nicknamed "Sven" when we worked together in the same foreign-car repair shop), had invited us to go motoring through India and Nepal. I explained to Mr Dinkel that this was the chance of a lifetime, as it is virtually impossible to rent a car in India, and that it isn't every day an Indian friend who actually *owns* a car calls up and says let's

drive across India to the Himalayas and see Mt Everest.

The Good Editor thought about it for a moment and said, "Sure, why not? Go ahead, I think. What kind of car will you be driving?"

"A 1966 Hindustan Ambassador, which is an Indian copy of a 1957 Morris Oxford sedan."

"Oh good. I don't think we've ever tested one of those."

By the time I left for India, my wife Barbara was glad to see me go. Two months of listening to old Ravi Shankar albums, mixing our own curry, reading the *Bhaga-*

vad Gita and listening to quotes from the Mahatma had just about done her in. By the end of February Barb began to worry that I would bring cattle home to wander freely in our yard and patio.

In March, Chris and I met in Chicago and hopped a British Airways flight to New Delhi, via London. Belted into the 747 at Heathrow, we were amazed to learn that our air route passed over the center of the Soviet Union, across Afghanistan, Pakistan and into India. Furthermore, we'd be going home through Kuwait, which at that time had 10 divisions of Iranian troops poised on its border. "If we live through this flight without becoming hostages or taking a SAM up our tailpipe," I told Chris, "I will devote my life to good works. Or buy us a drink."

It was a night flight, with the great darkness of Russia and Afghanistan relieved by occasional pinpoints of light below. We saw some kind of conflagration on the earth and tried to guess what it was. Oil field on fire? Kabul burning? Nuclear faux pas? A Pinto rear-ended by a Lada? No telling. When it comes to big fires, the Russians are a very private people.

Dawn arrived while we were still over Afghanistan, the sun rising on ridge after vast, endless ridge of lifeless, arid mountains that stretched through Pakistan and into the great desert of Rajasthan. "These people never water their lawns," I explained to Chris.

"What people?"

The first sign of trees and grass did not appear until we began our descent into New Delhi. The transition from desert

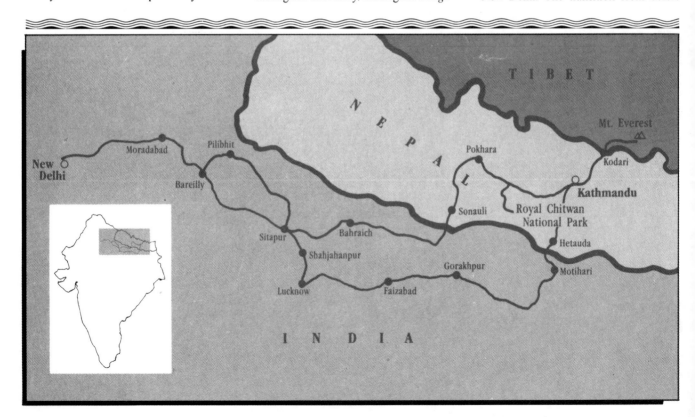

into the green fields of the Gangetic Plain is abrupt. You can see why the Aryans and Moguls were always pouring out of central Asia through the Khyber Pass to conquer India. They were looking for shade. What the British were doing there is not clear. Finding new markets for the Hillman Minx, no doubt.

Disembarking into a warm, sunny morning, we shuffled through customs in a pleasantly small airline terminal with songbirds flitting in and out of the rafters. Kirit was there to meet us as we stepped outside. He hadn't changed in the 10 years since we worked side by side as mechanics at Chris' shop in Wisconsin.

Kirit is a Sikh and is not exactly your typical Indian, if there is such a thing. He attended a New Delhi prep school, taught by Irish Catholic monks, and then went on to get a degree in mechanical engineering. Being a car buff, he came to the U.S. for automotive technical schooling and apprenticeship. After five years learning the trade, he returned to New Delhi and opened his own shop, which now employs 71 people.

Sikhs have been called "the Texans of India," a description that fits Kirit pretty well. He's about 6 ft 2 in., used to wear cowboy boots when he was in the U.S. and spent as much time as possible on hunting and fishing trips in Montana and Wyoming (his father was a tiger hunter of some reputation in India). He has an amazing ear for dialect and can easily drop his clipped Indian/English accent for a perfect Western twang or Southern drawl. He is the sort of person who can stride, resplendent in beard and turban, into the Sportsman's Bar in a small Midwestern town, order a beer, look up and down the bar and break the stunned silence by saying, in perfect John Wayne delivery, "Anybody around here hunt deer?" In five minutes he's made friends with every person in the bar.

Kirit loaded our bags into a Mercedes sedan. "It's not my car," he said. "It belongs to a friend who's out of the country." Leaving the airport, he sped through a stop sign at a busy intersection, leaning on the horn. Chris said, "Wasn't that a stop sign?"

"Nobody stops for stop signs here," Kirit said. "There are no rules, really. You generally drive on the left and you stop for red lights in the city, but that's about it. The important thing is to use the horn all the time. It's not considered rude. People expect it."

Traffic on the boulevard into New Del-

hi was a mixture of 5-ton diesel trucks with the brand name Tata emblazoned on the hood, Hindustan Ambassadors like the one we'd soon be driving, Premier President sedans, which are an Indian-made replica of the old Fiat 1100D, Triumph Heralds and hundreds of 3-wheeled, 2-stroke taxis. Any chinks left in this flowing, honking chaos were filled by bicycles, motorcycles and confused sacred cows.

New Delhi, like Washington, DC, is a "designed" city, a national capital laid out by British and Indian architects, just to the south of old Delhi. Along its broad, tree-lined avenues are government buildings, embassies, palaces built by maharajas and middle to upper-class suburban housing. Kirit drove us to his house, a spacious 3-story structure looking out on a park with a Mogul tomb dead in the center of it, so to speak. There he introduced us to his wife, Kitty, who turned out to be as witty and charming as Kirit. Then we met their 3-year-old daughter Sukhmani, house manager Terlok, driver Dayal Singh, several women who help Kitty with the household and the man fixing the front-porch damage incurred in last year's anti-Sikh riots.

For the next three days Kirit and Kitty showed us the city and cooked wonderful Indian meals for us, served in a dining room with snarling tiger pelts on the four walls. Dayal Singh drove us to Humayan's Tomb and the Qutb Minar, a 12th-century mosque, famous for having in its courtyard a 24-ft iron pillar that has remained mysteriously rust-free for 1500 years. The people who made the running boards on my last Volkswagen should look into this.

Kirit took us to Surya (Sun) Garage, his repair shop, a gas station structure set well back from a busy boulevard, all shaded by huge trees. Nearly all the work is done outdoors, and the shop can do almost anything: brake jobs, engine rebuilds, interior renovation, repainting, etc. Behind the shop, eight or 10 mechanics were busily rebuilding engines, squatting on the ground on large sheets of cardboard. "I put in some benches," Kirit said, "but no one used them. The mechanics prefer to work on the ground."

You don't have to be in India very long to notice that the ratio of people to objects (cars, houses, streets, trains, etc) is very high, so each object has a lot of people attending to its welfare. Nothing is thrown away; rebuildability is highly valued. Kirit's shop was a case in point. Sparkplugs were cleaned and gapped, tires retreaded, batteries re-cored and so on. Recycling is an art form, and the level of workmanship is quite high. The government encourages this trend by restricting car production to a small handful of models of fixed design and placing a heavy tax on new-car purchases. The Hindustan Ambassador, for instance, has been in production for 30

years, nearly unchanged.

Just before our arrival, Kirit had turned a dozen or so of his mechanics loose on our Ambassador. In a couple of days they'd restored the car to new condition. It had a perfect coat of forest-green lacquer, new vinyl upholstery and a rebuilt drivetrain. Under the hood was a 1.8-liter 4-cylinder ohv engine—essentially a detuned version of the early 3-main MGB engine—with a single semi-downdraft SU carburetor. Kirit had added a few deluxe touches, like an air conditioner, 4-speed gearbox, radial tires (Apollo Sherpas) and a stereo cassette player, complete with a supply of country and western tapes.

We awoke, more or less, at four on a Friday morning to head off on our grand adventure. Kitty made us a picnic basket and a thermos of coffee; we loaded our luggage into the Ambassador, pulled out the choke, engaged the Indian Lucas-replica starter and the engine settled into a nice mellow low-compression tickover. Kirit selected 1st gear and the sedan motored into the dark streets of New Delhi.

A few miles away we picked up Param-Git Singh (all Sikhs are named Singh), a longtime friend and hunting pal of Kirit's, who, at our urging, had decided to come along. Param, who recently sold his chemical business, had some spare time on his hands. He and Kirit would go with us as far as Kathmandu, then fly home to New Delhi, leaving Chris and me to explore Nepal and northern India on our own.

Old Delhi was just stirring to life in the morning darkness. Smoke from cooking fires of cattle dung mingled with morning fog, and strange, spectral scenes appeared out of the haze: hundreds of paperboys ripping into bundles of newspapers at a roundabout; a crowd of men gathered around trucks, waiting for temporary hourly labor on work gangs; bicycles, lurid movie billboards, white cattle and the faces of bearded *sadhus*, the wandering wise men, all lighted by a crescent moon breaking through the fog. We drove through a hundred small pockets of smell—diesel, dung, spices, baking bread, petrochemicals, bananas, sewage, oranges, death and spring blossoms, one on top of the other.

Crossing the Yamuna River out of Delhi, Kirit said, "Don't take any pictures of bridges. India is very security conscious, and taking pictures around bridges is against the law." Armed soldiers guarding the bridge watched our car go by.

A red ball of sun rose over the highway as we approached the Ganges, the road crowded with groups of Hindu pilgrims walking from their villages to this holiest of rivers and returning with decorative containers of river water slung over their shoulders on yokes. The road was good, an asphalt surface of about the width and quality of a paved county road in the U.S. Only the 5-ton Tata trucks made it seem

TO EVEREST

narrow. Tatas own the road in India, and they barrel along more or less in the middle of the pavement, horns blazing, forcing cars, bicycles and other road clutter to dive for the shoulder.

Most of these trucks are quite festive, decorated with so many Hindu icons, beads, tinsel and cosmic symbols that they look like five tons of good luck charm coming down the road. Their back bumpers are painted with a variety of messages (always in English) such as GOOD-BYE, TA-TA, HORN PLEASE and BYE-BYE GOOD LUCK.

Driving through the Indian countryside is like watching a slide show where the slides go by too fast to be absorbed or make sense. On the road from Moradabad to Bareilly, Chris and I sat with our heads gimbaling around at The Sights, taking in such roadside attractions as a caravan of gypsies with leashed greyhounds tied to their wagons (the dogs are trained to hunt rabbits for food, Kirit said), an ancient steam locomotive chuffing through a level crossing in glorious billows of soot, the ruins of a raja's castle, vultures the size of winged German shepherds eating dead cattle, kids using long sticks to prod an even longer snake into the path of our car (missed) and truck drivers feeding stale bread to a roadside convention of monkeys. Unlike so much of the U.S., India constantly reminds you that animals other than humans inhabit the earth. Not that India is short of humans.

We stopped at a soft drink stand in a small village for a bottle of Limca, a milky-sweet lime-flavored drink, and found ourselves surrounded by curious people who kept a distance of about 10 ft and stared at our car, shoes, clothes, bot-

TO EVEREST

tles of Limca, etc. I made eye contact with one or two people, smiled and said, "Ram-ram" and "Namastey"(two forms of hello), but to no effect. Chris had similar luck. There was no response. Not a nod or a blink or a smile. Just a stare. In the larger cities and towns, people were generally talkative and outgoing, but in the villages another mood prevailed. There was nothing threatening in it, just a shy, passive quality.

Turning off the main highway, we drove southeast toward the border of Nepal on narrower but less crowded roads. This was truly rural northern India, flat agricultural land dotted by picturesque villages of earth and thatch that could not have changed much in the past thousand years. Only bicycles and the thumping of an occasional gasoline-powered irrigation pump hinted at the 20th century. No cars and very few trucks. In one village we slowed to let an oxcart cross the road in front of us. The driver twisted the ox's tail and the cart shot across the road.

"We call that an Oxmobile with Twist-Tail Drive," Kirit explained.

At the end of a 300-mile, 14-hour day, we drove at dusk into Bahraich, a town of about 20,000, in which we appeared to have the only car, lost in a sea of bicycle traffic and humanity. Param asked directions and found an inexpensive hotel. Four dollars for two rooms and morning tea. After a scotch and soda in our rooms, we hired a pair of bicycle-drawn rickshaws to take us to a restaurant recommended by our hotel clerk. Our 97-lb drivers, in a living testimonial to gear reduction, pedaled us off through the dark and narrow alleys of Bahraich.

If Hollywood tried to build Bahraich as a movie set for the next Indiana Jones adventure, they'd go broke. It's too complicated and just plain anarchistic to exist anywhere but in some acid-soaked dream—or in India. By the time we got to the restaurant, 20 minutes later, I wondered if I was losing my mind. From the back of the rickshaw I'd seen (I think) about a hundred sacred cows, six kinds of free-roaming zoo animals, Hindus lighting incense in shrines, veiled Moslems, Gypsy women with gold earrings, monkeys in cages, monkeys out of cages, a caravan of Rajasthanis traveling by camel, beggars, blind people, street musicians, untouchables scooping up dung, several distinct madmen, a couple of weddings and funerals, Brahmans having their heads shaved in a barber shop and merchants in a hundred candle-lit shops selling a thousand small items. There may have been a few jugglers and clowns in there, too. I can't remember.

Our back-street restaurant served up a terrific meal of *dal* (cooked lentils), *chapatti* (a fried flatbread, used to scoop up *dal* and other foods), okra and eggplant curry and tandoori chicken, cooked in clay ovens. The waiter also brought water, which Kirit told us not to drink. "You mustn't drink the water anywhere in India unless you know it's been boiled. We can handle it, but your stomachs won't be used to it. Stick to beer, tea and bottled soft drinks."

Another surreal rickshaw ride brought us back to the hotel. Chris and I took a long walk through the city, which promptly closed down at 9:30, then returned to the hotel. Kirit and Param had picked an inexpensive hotel to give us a little slice of rural reality. And, while the place was serviceable, it was not easily mistaken for the Ritz. Our room had two beds and no windows or vents of any kind, but did have a slow ceiling fan whose purpose seemed to be to confuse the swarm of mosquitoes that rose to meet us when we opened the door. Wood paddles of the overhead fan slapped the air, reminding me of the helicopter sound track in *Apocalypse Now*. I fell asleep, dreaming I was in Vietnam again. During the night, Chris got sick.

Tea was delivered to our rooms by a small boy at six in the morning. Param had tipped (bribed) about a dozen people to make this early delivery possible—everyone from the hotel manager to the assistant doorman in charge of light switches. Chris was shaky, but feeling better, despite the sudden loss of 10 or 20 lb.

That morning we drove into another lovely, misty sunrise toward the border of Nepal. Kirit put a Ravi Shankar tape into the tape deck and explained that the music was a morning *raga*, starting quietly and slowly and building in complexity. It was the perfect music for a sunrise in India. Later he put on a Tammy Wynette tape and we all sang "Stand by Your Man" at the top of our lungs. By mid-morning we were in an area of forest and sugar cane fields.

"This is tiger country," Kirit said. "My father and his friends used to hunt tiger in these forests." He told us that a tiger can kill one person without being labeled a man-eater. If it kills two, it is destroyed or captured for a zoo. Several people had been killed by tigers in recent months, including an ornithologist who had wandered no more than 30 or 40 yards away from his car to observe some nesting birds. "Not much left but his hat and binoculars," Param added.

I looked into the surrounding forest with a new sense of awe. At road stops there would be no more wandering behind a bush for *this* cowboy.

We passed a villager herding some cattle through a clearing in the woods. "Good plan," Chris said. "Always travel with bait."

Early in the afternoon we approached the border crossing into Nepal, at the small town of Sunauli. An oriental-style portal and several customs and police buildings marked the border, which was busy with trucks, animal carts and a handful of German backpackers. An Indian police official invited us to get out of our car and sit at a long wooden table on the veranda of the customs house, where we filled out many forms. He looked at our passports for some time, then at our faces. How, he inquired, did two Americans and two Indians happen to be traveling together from New Delhi to Nepal?

"Well," I explained, "I'm an American journalist from California doing a travel story for a car magazine and Mr Beebe owns a car-repair shop in Wisconsin where Kirit Singh and I used to work in his employ. Mr Singh owns a car-repair shop in New Delhi now. We are old friends, taking a trip together, and Kirit's friend, Param-Git Singh has come with us. We are on our way to Kathmandu."

The policeman looked at my face, searchingly. Then he collected our passports and went into the office. He set them on his desk, made some tea, smoked some cigarettes and read through some forms. We waited in the heat on the veranda, sweat making armpit rings on our shirts. After about 15 minutes of waiting, I said to Kirit, "What's going on? Doesn't he believe my story?"

"I'm sure he believes your story," Kirit said. "It's too crazy for anyone to have made it up. I think he's just wasting our time because Param and I are Sikhs."

At last the man stamped our passports and let us go. We drove a few hundred feet to Nepalese passport control, filled out more forms, answered more questions, and then a soldier raised a crossing arm and let us into Nepal. As we drove away from the border station, a man stepped out of a small wooden shack near the road and waved half-heartedly.

"What does that guy want?"

"Who knows. Let's get out of here before we have to fill out another form." We kept driving.

This, we would later learn, was a big mistake.

The Ambassador rumbled straight north through hot tropical river-basin country. I'd expected a magical transformation at the border of Nepal, from farmland to majestic Himalayas, from cane fields and mosquitoes to glaciers and Gurkhas, but so far the country was flat as Kansas. I peered through the windshield into the humid afternoon haze. "It looks like rain up ahead," I said. "Look at that huge bank of clouds."

"Those aren't clouds," Kirit said.

THE ROAD TO EVEREST

New Delhi to Kathmandu, and back, via Oxford

PART II

BY PETER EGAN
PHOTOS BY THE AUTHOR

If you read Part I of this story last month, you may recall that our four traveling companions were about to begin an assault on the Himalayas in a 1966 Hindustan Ambassador sedan, a feat never before attempted, or at least never before mentioned. The Ambassador is an Indian-made copy of a 1957 Morris Oxford, an English car with four doors and three main bearings.

At the end of our last installment, the two Americans, Egan and Beebe, and their Sikh friends from New Delhi, Kirit Singh and Param-Git Singh, had just crossed from India into Nepal and were steaming northward through forested tiger country, ever closer to the high mountains and their ultimate goal—Everest. We continue:

WHEN THE HIMALAYAN foothills first loomed out of the clouds, they caught us by surprise. We'd been looking too low and too far away, and the mountains arrived more abruptly and much taller than expected. It was like turning to answer an insult in a bar and finding yourself staring straight into someone's belt buckle. The foothills were big.

"If these be the green fields, what be the brown?" I mumbled, possibly quoting some poet from college while casting a nervous eye on the water-temp gauge.

I shifted the Ambassador into 3rd, and we began climbing up a broad river valley into that part of Nepal that looks, on the map, as though someone had tried unsuccessfully to flatten out a piece of crumpled wrapping paper; vastly mountainous with only a few hundred miles of paved road.

The road itself was in excellent shape, following the path of least resistance along the rivers like a railroad line. But here and there the highway suddenly snaked steeply over mountain passes of 4000- to 5000-ft elevation, dropping us down into a succession of spectacular val-

TO EVEREST

leys. The Ambassador handled these 2nd- and 3rd-gear climbs without any visible strain; no overheating, rod knocking, pinging or other protest.

Had the Indians learned something about building cars that the British missed? It was certainly possible. Everyone else had.

On crossing the border into Nepal, the road traffic had jumped ahead about 20 years in design. The Hindustans, Premiers and Enfield motorbikes of India gave way to half a dozen brands of current Japanese cars and motorcycles, mixed with a fair number of Land Rovers and Toyota Land Cruisers. Driving in India, we felt there was nothing to suggest we hadn't just crashed our time machine in the mid-Fifties; in Nepal the traffic was very Eighties, except for a few Chinese-made trucks with a distinctly Korean War look about them.

The people, too, were different; rounder faces, Mongolian cheekbones and eyes, a generally stockier and more muscular build. And the mood also changed. Rural Indians tended to be rather shy and reserved, while in Nepal there was a sort of back-slapping spirit of cheerfulness that we felt immediately on entering the country. Crossing from India to Nepal was a little like leaving Maine and arriving in Mexico for Cinco de Mayo.

In fact, there was a Nepalese fiesta called *Holi* going on the day we arrived, apparently an auspicious time for weddings, parties, music and drinking rum. As we climbed through the green terraced hills at sunset, brightly costumed throngs were converging on the villages. By darkness, every village had a roaring bonfire surrounded by a wild revelry of singing, drinking, shouting, tambourines, drums and horns, all sounding exactly as though someone had spiked the punch at a Salvation Army street-band convention.

Nepal was beautiful, even after sunset. The hills were dotted with thatched-roof huts of wood and brick that clung to impossibly steep slopes, reachable only by rickety suspension bridges or near-vertical footpaths. At night these remote houses were lighted by candles and oil lamps that gave off a warm orange glow and made them look like jack-o'-lanterns scattered off through the hills.

We cruised late at night into Pokhara, one of the few towns in Nepal large enough—and flat enough—to have an airport. Being in the mood for hot showers, we checked into the relatively lavish Crystal Hotel, the sort of place that accepts American Express and has things like room service and doormen. And doors.

The Pokhara Valley is famous for its panoramic view of Annapurna and the unclimbed Machhapuchhare, or "The Fishtail," as it's called. Unfortunately the mountains were obscured by clouds when we woke up in the morning so we bought post cards of the view and looked at those during breakfast.

"I'll bet this is a beautiful place," Chris said, examining his cards.

"Yes, it is," the waiter assured us.

After changing our Indian rupees to Nepalese rupees at the hotel, we filled up with gas (21 mpg, $2.20 per gallon) and set out along the Seti Khola River Valley, eastward toward Kathmandu. The scenery along this river is breathtaking, its deepest gorges spanned by the sort of suspension footbridge the Camel Filters guy is always repairing; the kind that used to spill half a dozen porters, complete with luggage, in every Tarzan movie.

At lunchtime we stopped in the little town of Dhumre at a spot called the "Mustang Lodge." Lunch was the usual Indian/Nepalese fare; curried chicken and vegetables served with *dal* and *chapatti* (lentil mixture and unleavened bread Frizbees) and Star Beer—made in Hetauda, Nepal, "Under German Technical Collaboration." About a hundred children gathered around our table, apparently to see how foreigners eat.

The restaurant was a board building with a high stone sidewalk in front, like something from a Nevada gold mining camp. With horses and mules in the street, these towns we drove through felt, in fact, very much like a part of the Old West. Even the character of the people was right. There's a freewheeling, woolly optimism in the hill people that would make Teddy Roosevelt and Buffalo Bill feel right at home. Except at toll stations.

We'd hoped to enter Kathmandu in daylight, but it was not to be. Nepal is full of little toll stations where every truck and passenger-car driver has to stop to pay about 5¢ and fill out a form that makes the IRS 1040 look like a traveler's check, recording such information as date of birth, engine serial number, favorite color, etc. This is all written in a Book of Kells-size volume that, when full, is put out to molder in a back room. All very official, with soldiers looking on.

To make matters worse, we were supposed to have bought something called a Road Permit at the border. At every toll stop we were asked to produce this mysterious document, which, of course, we did not have. Our missing Road Permit created great confusion among the soldiers and toll officers. "But you must have it," they would say.

"Well, we don't," I'd answer. "Nobody told us we needed one at the border."

"But you must have it! You cannot be here without one."

"Well, we don't have one . . . and here we are."

This interesting conversation would go on for four or five minutes, and then they'd make us fill out a form (chassis number, last book read, hat size, etc) and reluctantly let us go.

"Do you remember when that little guy came out of that shack and waved to us right after we crossed the border and we kept driving?" Chris asked.

"Yeah."

"I think he may have wanted to sell us a Road Permit."

These road stops propelled us into Kathmandu at about 11:00 p.m. After searching for a hotel with just the right ambiance—not too modern and touristy, but without swamp creatures in the shower—we checked into the Mt Makalu Hotel, a modest but clean little 5-story place on a side street.

I was a little disappointed driving into Kathmandu. Reading *National Geographic* all these years, I'd expected a cross between a mountaintop monastery and Xanadu. Instead, we drove into the outskirts of a dusty, noisy city of 400,000 with apartment buildings and broad avenues that felt more like the seedier suburbs of Paris than the forbidden city of a mountain kingdom.

But in the morning light we took a walk and found that our hotel was literally just around the corner from the edge of the Real Kathmandu, the city that was not accessible to foreign visitors until the road to India was completed in 1955, and not visited by any great number of tourists until the airport was built in 1968. Xanadu, it turned out, was only about a block away.

There are supposed to be 2000 temples and shrines in Kathmandu, and the number is believable. Every street in the old city seems to lead into a small square or plaza crowded with temples whose statuary and pagoda-shape rooftops overlap and overshadow one another. The smell of incense, the ringing of ceremonial bells, pilgrims, monks and sacred cows filled the squares and narrow alleyways. There were Western tourists here and there in the crowd, but the throngs of people in the streets were overwhelmingly Nepalese, so the exotic feel of the city was largely undiminished by tourism.

The city is in a broad valley only 4500 ft above sea level and, being on about the same latitude as Daytona Beach, is quite warm and balmy, even in March. The mountains that loomed to the north of the city were frozen and forbidding, but in Kathmandu you could wear short sleeves in the afternoon and a light sweater in the evening.

After two days of hiking around Kathmandu, we'd had enough of the good weather and decided to descend once more to the steamy Terai plains at the foot

TO EVEREST

PHOTO BY CHRIS BEEBE

of the mountains. After making reservations through a Kathmandu travel agent, we drove 100 miles down to Gaida Wildlife Camp and Jungle Lodge at the edge of Royal Chitwan National Park. We met a park guide at the end of a paved road near the park, and he advised us to leave the Ambassador and ride to the lodge in his Land Rover. Chris and I climbed into the Land Rover, while Kirit and Param decided to try out the Ambassador as a cross-country vehicle. It all went fine until they crossed a river and water hit the fan, breaking a blade. Kirit broke off an opposite fan blade for balance and made it to the lodge without further trouble.

Gaida proved to be a rustic but nicely built collection of huts and cottages in a picturesque spot on the Rapti River. We had a drink at some outdoor tables, rested awhile, took a quick lesson in elephant behavior and then went on a 2-hour elephant trek through the forests and along the river. Param and I paired up on one elephant, sharing a *houda*, or platform. Our *mahout* (driver) got us in close to a family of rhino and managed to scare up half a dozen varieties of deer and some wild pigs.

I'd never ridden on an elephant before and was amazed at the pounding, rough ride, a bumpy rhythm of starts and halts. Chris caught the mechanical essence of the motion. When we came back from the ride, he wrote in his journal that it was "like riding on stilts made of solid lead, propelled by silent motors gyrating forward and backward beneath the elephant's shoulders; a fluid yet immensely heavy movement." After that, I let Chris write all the notes.

That night we were having drinks around the campfire after dinner and laughing it up with some Germans when I suddenly became dizzy. I staggered off to bed and spent the night alternately shivering and sweating with chills and fever. By morning I felt fine. We decided I'd had either the 24-Hour Malaria, Evening Cholera or Temporary Black Death.

The (now) 4-blade Ambassador climbed back up to Kathmandu the next day without problem, and Chris and I dropped Kirit and Param off at the airport for their flight back to New Delhi. Chris and I were now on our own. So the next day we decided to drive to Tibet.

Our map showed paved road winding north to Tibet, with only the last 16 miles unpaved. We left in the morning and had our best view yet of the high Himalayas, on a ridge just outside Kathmandu at Dhuklikhel (rhymes with Phuklikhel). The closer you get to the high Himalayas, the deeper the river valley and the harder they are to see; so the peaks disappeared from view as we drove north.

We'd hoped to reach the Tibet border by late afternoon, but ran into a roadblock. A bus had left the road in a big way, landing on the riverbank below, and a cast of thousands had retrieved it with the clever use of blocks and tackles. It was now lying on its side across the road, and they were trying to move it aside to let traffic through. Chris and I watched this operation until sunset, then drove back to a village called Lama Sangu to look for a place to sleep.

There we found a small, unmarked inn with a restaurant of sorts at the rear.

A woman appeared and led us up a ladder/stairway to a large loft with five or six wooden cots. "Looks good," we said. Dinner was served by lantern light in a smokey little back room—excellent rice and *dal*, served with mysterious pieces of curried chicken not even the Colonel himself could identify. After dinner we went to bed in the loft and were joined by three truck drivers, an elderly woman and three children sleeping on straw mats, a married couple with a child, an old man with a cough that sounded like broken glass and some dogs. During the night, more truck drivers arrived and slept on the floor.

When we awoke in the morning, there were 18 people sleeping in the loft. No wakeup call was necessary as we had many alarm systems, including a roaring bus engine outside the window below, dogs barking, dogs fighting, death-rattle coughing from the old man, a radio playing Nepalese drum music downstairs, cooks clattering pans, cooks fighting, children feeling the hair on our Caucasian arms and thick smoke pouring up through the stairwell. Chris rolled over and said, "Either breakfast is ready or the kitchen's on fire."

I pulled a bedbug off Chris's back and we went down for breakfast. After a *chapatti* and tea breakfast, I paid our bill: 28 rupees for dinner, breakfast and a room for two, or about $1.27. Children followed us everywhere in Lama Sangu, and before we left Chris and I taught them to do the Charleston and snap their fingers in a horse-trotting rhythm, like Curly of the Three Stooges. When we drove out of town, the streets were full of kids weaving their knees and snapping their fingers, frowning with concentration.

I said to Chris, "Can't you imagine some cultural anthropologist coming into this village 20 years from now and trying to figure out what went wrong?"

We arrived at the bus accident just in time to see it tipped upright and out of the way, and continued toward Tibet. After driving through the border town of Kodari, we came to Friendship Bridge, beyond which there was no passage without a visa. Chinese trucks were parked on the other side, but there were no barricades or barbed wire. Just a couple of soldiers. We took a few snapshots and headed back.

Coming into Kathmandu late in the afternoon, we found the streets empty of traffic but lined with waving people, banners, flags and rose petals. Our dusty green Ambassador cruised through town like a 1-car parade, cops and soldiers waving frantically at us. "This can't all be for our benefit," I said. "Some Nepalese bigwig must be coming through. Let's get off this main drag and back to the hotel before someone shoots us."

A few minutes later a motorcade came down the street, and we found out it was the Queen of England. Having not seen a paper for over a week, we hadn't known she was in town.

TO EVEREST

The next day, we booked ourselves on an early morning mountain flight with Royal Nepal Airlines ($75 each) and drove out to the airport before sunrise. It was time to see Mt Everest.

We boarded a pressurized twin-engine plane with about 30 seats and took off into the rising sun. After circling Kathmandu for altitude, we climbed toward the wall of high snowy peaks that parallels the Nepal/Tibet border. We kept identifying tall mountains as Everest, but the stewardess said, "We aren't there yet. You'll know it when you see it."

And we did. It was the big mountain with the long plume of snow and vapor streaming straight off its lee side like an aviator's white silk scarf. Everest was as forbidding and majestic a thing as I've seen, and its immensity filled me with the real meaning of all the expeditions I'd read about since childhood. I breathed a quiet whistle of respect for the Hillarys and Tenzing Norkays of this world. It was almost wrong that we were seeing this peak from 20,000 ft so effortlessly. Only about 9000 ft below the summit, with no frozen toes, no oxygen bottles, no nights in a fluttering tent at Camp Five.

Then, as if to remind us where we were, wind began to shake the airplane around like a dog with an old slipper. We held on for five or 10 minutes of severe buffeting and the stewardess said, "This is the worst turbulence we've had all year. The weather is growing worse now, and this is the last mountain flight of the season."

When it calmed down a bit, I took some pictures through the icy window of the airplane, and a man in the seat ahead of us said, "You have a better view from this window, if you want to try it. You can see past the wing."

Chris and I looked at the man and then at each other. Chris looked back at the man and said, "Are you Neil Sedaka?"

"Yes," he said, "I am."

We traded cameras and windows and talked as the plane circled back for another pass at Everest. Sedaka was on a world tour and had taken a side trip to Kathmandu. He seemed a very nice man, friendly and down-to-earth.

When the plane landed at Kathmandu, we said goodbye and Chris and I drove to the hotel harmonizing on "Calendar

PHOTO BY CHRIS BEEBE

PHOTO BY CHRIS BEEBE

Girl" and "Breaking Up is Hard to Do."

Late that morning we checked out of the hotel, adjusted the valves on the Ambassador using a Gaida Game Park business card for a feeler gauge (estimated thickness, 0.015 in.) and loaded our luggage. The starter motor wheezed into engagement and feebly turned the engine over. Chris turned to me and said, "How's this for a new bumpersticker: IF LUCAS MADE GUNS, WARS WOULD NEVER START."

But the engine fired, and on the way out of town we stopped at a store to stock up on bottled mineral water for the drive across northern India. We bought Golden Eagle Brand Mineral Water, a product of India. "Must be good," I said, reading the label. "It says it's from Himalayan Springs."

"So is the Ganges," replied Chris, whose continued stomach trouble had made him cynical of many Eastern food products.

The road from Kathmandu south to India was by far the most beautiful and the most tortuous we'd traveled in Nepal. It was also remarkably free of truck traffic.

This was the ancient trade route from India to Kathmandu, the path over which pianos and Fords and chandeliers from Paris had been carried on the backs of men, or—more likely in Nepal—women. (The road was full of women struggling under gigantic loads on their heads and backs, while the menfolk walked unladen a few steps behind, grinning cheerfully and smoking cigarettes.)

Progress was slow. We were still in Nepal when darkness fell, negotiating mile after mile of downhill switchbacks. Chris took over at the wheel, and as we drove late into the night, I could tell from his disconnected speech, or lack of it, that he was getting very tired. So I tried my Test Question on him: "Didn't you say your sister was a door gunner for the First Cav at Binh Dinh?"

When he agreed, I knew it was time to stop. We found a nice old motel called the Avocado near the town of Hetauda. In the morning we left the mountains for a descent into the Terai Plain.

We crossed the border at Birganj. Rather, we tried to cross. We still had no Road Permit, so they held us in a customs hut all day while various customs agents tried to call the border station at Sonauli to find out why we hadn't been issued a Road Permit. As none of the telephones worked, this was a difficult task. Every hour, they shouted, "HELLO! HELLO! HELLO!" into the phone for 10 minutes, then gave up. Late in the afternoon they said everything was okay and let us go. We drove into India.

One hundred miles later our highway turned into a dirt road and then into a goat path that meandered aimlessly through villages. Between our "Single-A" road map of India, and the signs being written in Sanskrit, we were soon lost and getting alarmingly low on gas. After a few hours of bumping and trundling through a few hundred villages, we came to a broad, shallow river and followed its bank south in search of a bridge and a real road, which we found at sundown. Four hours later, exhausted, we cruised into the city of Gorakhpur checking into what we would later call The Worst Hotel in the World. Unequivocally.

It took us about an hour to check in, as the desk clerk had to make sure that every number and letter was legible and consistent on all four lengthy registration forms. While we filled these out, a man in the hallway was trying to place a telephone call. The conversation went roughly as follows: "HELLO! HELLO! HELLO! HELLO! HELLO! HELLO! HELLO! HELLO! HELLO!" and so on, for about an hour, each HELLO! shouted into the mouthpiece at 5-second intervals.

"Maybe he's talking to the guy at the border," Chris suggested.

"Reagan must have deregulated their phone system."

At midnight we finally got a key to our room. It had two beds, fearless roving gangs of roaches, a dense cloud of mosquitoes, electric-blue walls, no towels or toilet paper, one fluorescent light and two drinking glasses half filled with brown water, drowned insects and cigarette butts. We dropped our luggage and took a look around.

"I think they used this sheet to wipe the bathroom floor," Chris said, examining the one sheet on his bed.

"No chance of that," I said, looking in the bathroom. "No one has ever wiped this floor. With anything. Ever."

It was St Patrick's Day and we wanted to have a drink to celebrate, so I walked down the dark hallway, tripping over some garbage, to ask the desk clerk for some clean glasses.

"But you have glasses in your room."

"Yes, but they have cigarette butts and dead flies and brown water in them."

He looked at me as though I were out of my mind, trying to discern the problem.

Finally he clapped his hands and a raggedy, bag-of-bones apparition of a man unfolded himself from where he'd been squatting in a dark corner. The hotel clerk shouted at him. The man nodded. Then he shuffled down to our room; picked up the offending glasses; opened our shutters; tossed the dirty water, dead insects and cigarette butts into the street; and put the glasses back on our night stand. He folded his hands, bowed slightly and left.

Chris and I stared at the glasses for a moment, our brains humming with fatigue, and then we started to laugh. We fell back on our beds and laughed for about five minutes, out of control. Chris finally caught his breath and said, "Pete, we've found it. After all these trips, we've finally bottomed out. This is The Worst Hotel in the World."

"Now I can die happy."

Recovering our senses after a time, we sat up and passed a bottle of Scotch back and forth wishing each other a happy St Patrick's Day.

After we went to bed, a terrific dog fight broke out just beneath our window. These occurred nearly every night on our trip, but this one was worse than usual, so we went to the window and opened the shutters. Twenty-five or 30 dogs had a sacred cow cornered against our hotel and were attacking it like a pack of wolves. The cow lowered its horns and charged the dogs, then suddenly bolted down a dark alley. Many dogs followed, but most remained beneath our window and scrapped among themselves. We closed the shutters. "I love the gentle spirituality of the East," I said, "and the deep veneration for life."

Chris shook his head. "Think of the vet bills in the morning."

We awoke early because there was someone vomiting in the stairwell below our room and someone else on the hallway phone shouting, "HELLO! HELLO! HELLO! HELLO!" So we got up and left at dawn. Chris and I had both lost our appetites for regular meals, for some reason, and had taken to buying bananas and oranges at roadside stands and eating nothing else. We'd both lost about 10 pounds. We loaded up on fruit and headed out of Gorakhpur.

Looking at the map, it appeared we could make it back to New Delhi in one very long day's drive, so we decided to keep driving until we got there. We traded stints at the wheel, ate oranges and watched the lovely/strange/sad/exotic Indian countryside roll by.

Through Basti we drove, across the Ghaghara River into the ancient city of Faizabad, full of spectacular temples and mosques and giant shade trees filled with monkeys; through the crowded, endless bazaar of Lucknow, braking, honking and swerving to miss a thousand bicycles, pedestrians, trucks and oxcarts (Chris counted 126 honks of the horn in 5 minutes of Lucknow traffic); past endless broken-down trucks on the highway, always with bad axle bearings and rear axles slid half-way out into the flow of traffic; into the night west of Sitapur, stopping for half a dozen level railway crossings where we waited with oxen and rickshaws for an old English-made steam locomotive to come roaring by, showering magnificent sparks into the darkness and red boiler glow onto the men shoveling coal. (We waited half an hour at one crossing after the train had gone by, the man in the crossing hut shouting, "HELLO! HELLO!" into the phone, trying to get permission to open the gate from some distant, unreachable authority.)

At Shahjahanpur a kindly gas station owner, pleased to exercise his English, served us tea, while one of his employees filled our tank. With diesel fuel. Which we discovered a few miles from town, when the car would run only on half choke. We labored on through the night on 50/50 diesel and gasoline, through Bareilly, where we almost hit an elephant as it stepped off a street corner (no taillights, Chris said) and into an apocalyptic thunderstorm and fork-lightning show between Rampur and Moradabad, swerving, braking, downshifting, diving off the road and missing by inches streams of Tata trucks with their headlights mercilessly on high beam, sharing the shoulder with every known form of shadowy conveyance with no lights at all, past truck stops with outdoor cots for drivers, their all-night food-kitchen fires flickering in the dark and radios blaring the wailing, slippery notes of Indian music, and still more broken trucks, their drivers repairing axle bearings by the light of campfires built on the road, under their own trucks. We ate our oranges and drove on.

At one in the morning we were adrift in the vast eastern suburbs of New Delhi, searching muddy roads for a bridge across the Yamuna River. By 2:00 a.m. we'd found Connaught Place, but still couldn't find the avenue to Kirit's house. So we hired a 3-wheeled cycle cab to find the address. I rode in the cab and Chris followed in the car, on a wild, careening 70-mph trip through the deserted boulevards and roundabouts of New Delhi. At 3:15 a.m. we were in Kirit and Kitty Singh's driveway. Home safe.

Over the next two days, Kirit and Kitty took us out to superb restaurants for dinner, Param-Git threw us a party at his beautiful house at the edge of the city and we got to meet Kirit's mother, a sparkling, intelligent woman who told of her family's flight from Pakistan during the partition of India.

Kirit turned our steadfast Hindustan Ambassador over to his mechanics and had it washed, tuned, diesel-drained and the road dents pounded out of the oil pan. He said, "If you guys want to use the car to see a little more of New Delhi before you leave, feel free."

We both declined. Chris said, "We made it through Nepal, across India and back to your house without an accident. We've used up all our luck. We can't drive here any more."

At the airport we said goodbye to our good friends promising to see them again, on one side of the world or the other.

On the flight home, Chris and I agreed that it had been a fine trip and a great adventure and that we would never do it again for a million dollars.

Nor would we advise anyone else to take an automobile trip through India and Nepal.

There had been too many opportunities to hit a cyclist, injure a running child, collide with an oxen, run head-on into the front of a 5-ton truck, to injure or be injured days and miles and medical light-years from a working telephone, a hospital, electric lights, a lawyer, an airline terminal or a sympathetic cop, soldier or bureaucrat. We'd missed and been missed by disaster by too few inches too many thousand times. We'd gone a little too far out on the limb and knew we could never make the trip again without having something go terribly wrong. In the brilliant clear light of retrospect, a car trip through India and Nepal was not really such a good idea.

"Next time," Chris said, as the green coast of India gave way to the placid blue Arabian Sea, "I think we should take motorcycles."

"I was thinking the same thing," I said. "Safer. Not so wide. A couple of India Enfield 350s would be perfect." ◎

MODEL A ODYSSEY
Part 1

MODEL A ODYSSEY

Part 1

*A ballad of
foggy mountain breakdowns,
Oklahoma hills
and that ribbon of highway,
played in four-five time,
hummed in the key of A*

BY PETER EGAN

PHOTOS BY THE AUTHOR

AT THE STATE Line Steak House & Truck Stop in Thayer, Missouri, a man at the counter asked where we were going in that old Model A Ford. "From Wisconsin to California," I said.

He looked out the window at our car thoughtfully, his toothpick migrating slowly from one side of his mouth to the other. "Good car," he said. "Bad map."

"We're going a little out of our way," I explained. "We've never seen the Ozarks or Texas, so we're taking the southwest route on 2-lane roads, staying off the Interstates."

"Where are you headed next?"

"Across north Arkansas to Okemah, Oklahoma, then south into Texas."

"Why Okemah?"

"It's Woody Guthrie's home town."

"Well, I expect you can make it from there to California in a Model A. That's how all the Okies got there."

Yes, I thought, *but even the poorest dust-bowl Okies didn't drive cars that were 57 years old. On the other hand, they*

didn't have credit cards, so maybe the score was even.

To be fair about it, we had a few other unfair advantages, too. Like six months of preparation and an $8500 budget to buy and rebuild the car. And God knows it had needed some rebuilding. My friend Chris Beebe and one of his mechanics, Bill Putnam, had been working on the Model A nearly all winter at Foreign Car Specialists, Chris's repair shop in Madison, Wisconsin.

The crazy notion of buying this car had hit us like a double-pronged pitchfork of lightning back in the autumn of 1986 when I was visiting Madison. Chris and I were driving from his farm to the city one morning when we saw a Model A Ford sitting on the lawn of a farm house. The car had a "4 sale" sign in a window. Naturally, we stopped for a quick look.

The car, a Tudor sedan, was dark blue with black trim and looked quite noble and upright in the early morning Octo-

ber sunlight, like an echo of the gothic Midwest farmhouse behind it. "That's a handsome car," I said to Chris. "Have you ever driven one?"

"No."

"I haven't either."

We pondered that stark reality for a moment; both of us involved with cars all our lives and we hadn't driven one of the most famous cars ever built, reputed to be the most commonly restored classic in the world.

"We should do something about that," I said.

Upon returning to California, I dived into the R&T library for a little basic Model A research. It was a car I really knew very little about. Born in 1948, I was in that generation that was just a little too young to have taken advantage of the cheap, throwaway Model A as basic youth transport. By the time I achieved car lust, they'd already become minor classics, antiques owned and driven by the kind of older person who collected

grandfather clocks and wall barometers. Guys my age wanted '51 Mercs, or any cheap car with a V-8 in it. James Dean did not drive to school in a Model A. And I didn't want one either.

That sense of the car's antiquity was reinforced when I later took engine-rebuilding courses at the county technical school. The Model A guys were a breed apart, usually in their 50s and 60s, filled with folksy stories about the glory days of pouring babbit, bailing-wire repairs, cylinders overbored through dozens of oversizes and, of course, the classic about using a piece of your leather belt for a rod bearing until you made it to the next town. There seemed to be a dangerous level of tooth-sucking going on around these cars, and, anyway, I was too busy turning a Sprite racing engine into a 948-cc fragmentation grenade to pay much attention to the virtues of these agricultural engines. They just

weren't my kind of machine.

So it's hard to say what chemistry or play of sunlight on metal caught my attention as we passed that Model A in the farmer's yard. Maybe I was getting old. Or maybe it was shopping malls and MTV and throwaway econobox cars that were getting old. Whatever the cause, the Model A suddenly struck me as looking very right, lean and flinty as Henry Ford himself, honest and mechanically direct, and I felt the time had come to drive one.

Doing my homework back at the office, I learned that the Model A was built from 1928 to 1932 and that Henry Ford didn't even want to build the car, feeling that the Model T was all America really needed. America didn't agree, however, and was buying more Chevys and the like, and ever fewer Model Ts. When Henry finally got behind the "New Ford," he put everything into it and released the car to a public nearly delirious

with curiosity, amid hoopla unequaled since the Lindbergh reception. The first cars were carefully doled out to the influential and famous, like Thomas Edison, Will Rogers and Mary Pickford. It was a credit to Henry's mechanical instincts and his son Edsel's fine eye for design that one of the cheapest cars in America was sought after by people who could afford much more than the $495 base price.

Later, as the Depression set in, dust-bowl farmers added another, more somber, side to the car's place in history by loading roofs and running boards with furniture, children and crates of chickens and striking out for the Promised Land. By the time the Model A went out of production in 1932, it was as firmly entrenched in the American consciousness as wall phones and Main Street.

And just half a century or so later I sat in the R&T library, looking at picture books of the car and trying to decide which model looked best. It was hard to pick a favorite; they were all remarkably well styled cars. The coupes, roadsters and cabrios all looked jaunty as a tweed cap and spats; the phaetons were long

and elegant; the Tudor and Fordor sedans practical and roomy, with lots of side window. Not a dog in the lot. Any one of them would do. Even a Tudor sedan sitting back on a Wisconsin farm.

After getting Editor John Dinkel's blessing (and promise of funding), I called Chris back and asked how he felt about driving a Model A Ford across the country.

"I've been looking at that car every day on my way to work," he confessed. "There's lots of room for luggage in the back seat. We've never had room for luggage before on a car trip."

"I suppose a guy could carry, oh, say, a couple of guitars back there."

"I was just thinking the same thing."

"I think Woody might approve."

By the time we sprang into action, the Model A in the farmyard had been pulled off the market by its vacillating owner. Chris found another one, however, on a nearby farm. An all-black 1930 Tudor Sedan, recently restored, except for its stripped-out interior. He bought it for $3800 and drove it off to the upholstery shop for a $1500 upholstery job. I called and asked him how it drove.

"Pete, I hate to say it, but it's the worst car I've ever driven. It changes lanes by itself, the brake pedal bottoms out on the engine block and the engine vibrates through the steering wheel so much my arms are numb all the way up to my shoulders. Everything jitters and rattles. I can't imagine driving it all the way across the country. It's so bad I'm almost speechless."

"Great. Sounds like we got a winner there. Can you fix it?"

"Give me a few months."

All that winter I called Chris about every 15 minutes to see how the car was coming along. On one call, he said, "You know how they say anyone can work on a Model A? Well on this car everyone has. And not a single one knew what he was doing. The brakes are assembled wrong, the steering box is a mess and our 'newly rebuilt' engine has a rod installed backwards and chunks of babbit in the oil pan."

But just six months later, all was healed. Bill Putnam and Chris had rebuilt the steering, installed a new engine, new wheels and tires, brakes, wiring, etc. By the beginning of May, $3000 later,

Model A Odyssey
Part 1

Above: Ozark mountain breakdown, cured with a little help from our friends. Left: Fast food joint, Arkansas style.

the car was ready, so I flew in from California to take advantage of all their hard work. We washed and waxed the Ford and that evening headed south out of Madison toward Chris's farm. My first driving lesson.

"To start," Chris instructed, "you turn on the gas under the cowl tank, retard the spark with the lever on the left, give it a little throttle with the lever on the right, pull out the choke/mixture knob, push the starter button with your foot and let go of the choke knob before the engine floods. Once it starts, it runs best on full advance."

I sat at the wheel like a cave man who'd just discovered a vending machine in the forest. Too many choices, a dim brain and no spare change. But gradually it all sank in, I moved all the right buttons and levers and the car started. Its 200-cu.-in. flathead-4 idled with a hollow doopadoopadoopadoopa, sounding suspiciously like the engine that used to run the Ferris wheel at the county fair.

Driving time: A 3-speed floor shift, with reverse at the upper left of the H. Ease it back into 1st with the slightest graunch, let out the very normal clutch and you're off. Steering is heavy, driver's seat close to the wheel for good reason. The shift linkage is worn, Chris says, so shift to 2nd with a slow double-clutch, and into 3rd with a slow count to four. It works for him, but I announce to the whole world that a shift of sorts has taken place, one that won't soon be forgotten. "It takes practice," Chris says charitably. 'Deed it does.

The pre-hydraulic mechanical brakes work, kind of, but not equally all at once, and you can feel the stretch of metal rods and the collective slack of fulcrums and pins. Beyond a certain point, the more you push the less you stop. There'll be no tailgating in this baby. There's a lot of buzz and rattle as you build speed with the A, but between 35 and 45 mph it settles into a friendly, relaxed cruise mode, chuffing along happily. Model A folks have told us not to cruise faster than about 47 mph if we want our rod bearings to arrive at the Coast with the rest of the car. That seems like good advice. Fifty feels slightly stressed and hectic.

Handling? Not so's you'd notice.

Steering, though, is remarkably tight and accurate—better than the MG-TC Chris and I drove to Road Atlanta a few years ago. All Chris's and Bill's labor on the steering bits had paid off. Cornering, however, is of sub-TC quality. Weight really loads up on the outside front tire and you feel the car leaning hard on those tall 19-in. wheels, like Abe Lincoln rounding a sharp corner on stilts. A sports car it's not.

> *A perfect platform for viewing that part of America still innocent of the exit ramp and the frontage road.*

On a Sunday morning we loaded the car for travel. Tools, a few ignition spares, camping gear, guitar cases and suitcases. In keeping with the spirit of the trip, Chris had come up with a couple of ancient, battered suitcases that looked like something Ma Kettle might have put out for a garage sale. He also produced an old violin case, on the theory that no Model A should be without a violin case for luggage.

"What's in there?" I asked.

"My shaving kit."

Loaded down with luggage, two spare tires and Chris's famous "Rand O'Mally," the world's most outdated road atlas (Arizona *Territory?*), we hit the road. Full advance.

Country roads took us chugging through the hilly farmland of southern Wisconsin, across the Mississippi into Iowa, Land of Storm Opportunities. (Every time we've driven into this

state, we've hit a storm just across the border. I don't know why they don't do something.) We rocked through a gusty tree-ripping thunderstorm and I said to Chris, "Maybe we'll see a tornado on this trip. It's the right season, and we'll be driving right through Tornado Alley. I've always wanted to see one."

Chris grinned and shook his head.

"It's the product of living in California too long," I explained, "with that shopping-mall climate. I could use a little excitement."

We drove out of the storm into a beautifully clear, benign evening at Iowa City. The Model A was not happy in Iowa City—or any town with more than one stoplight—but it liked the rural 2-lane blacktop. With the big side windows down and the bottom of the windshield tilted outward on a warm day, it felt as open and cool as a shaded porch swing. A perfect platform for viewing that part of America still innocent of the exit ramp and the frontage road.

In the morning we jogged south and stopped for lunch in the little town of Douds, Iowa. We'd barely stepped out of the car when the postmaster, Mr Wendell Peacock, came out of the post office and greeted us, admiring the car and offering the water we needed to top up the radiator. We walked into the Douds Cafe and the waitress paused while pouring someone's coffee and said, "Is that a '30 or a '31?"

"It's a '30," I said.

Clatter and conversation resumed, bets won and lost. "I thought so," the waitress said, throwing a knowing glance at a back booth. "I learned to drive on a '30. My dad bought it for $500, *new,* and it went chuggata-chuggata, right down the road." Half the cafe, mostly men in seed caps, emptied out to look at the car, then returned to lunch. After that, the cross fire of conversation in the cafe was thick with old-car words: ". . . hand throttle . . . babbit . . . busted brake rods . . . hand crank . . . "

As we drove south into Missouri, it

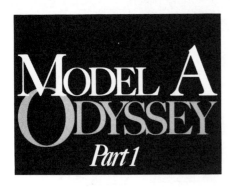

MODEL A ODYSSEY
Part 1

Left: Royal Motel, Norfork Lake, Ark. Above: A small part of the Bill Dean Collection of fine used Ford parts.

happened over and over again. It seemed nearly everyone over the age of 50 had learned to drive in a Model A or had owned one—or half a dozen—and had never paid more than $100 for a used one, $35 being a more common price. We also learned that if you like to meet people, there's no better calling card than a Model A. It's accepted immediately as a friendly, non-threatening device, a car that ordinary people could afford when it was new and can still afford, if they want one. When Chris and I drove an MG-TC across the country, people were interested and helpful, but more reserved with their goodwill until they checked us out. When we stepped out of the Model A, they automatically assumed we were All Right and walked right up to talk to us. After all these years, the old Ford is still a car that belongs, wherever you happen to stop. Its possession puts no distance between you and anyone you meet. It is an oddly potent reminder of good things to a great many people.

We camped in central Missouri at Graham Cave State Park, setting up our tent by the light of the moon and about 10 billion fireflies that swarmed the nearby meadow like a luminous carpet. We slept, more or less, through a thunderstorm and woke to a clear morning of car maintenance, adjusting wheel bearings, brakes and one noisy valve. The car had already crossed two state lines, running without complaint, and the silence was making us a little edgy.

Not for long, though. As we cruised south through that lovely green, complex, convoluted mass of hills, valleys, nooks, crannies and forests called the Ozarks, the engine began sounding overly rich. We stopped at a gas station and found fuel pouring out the top of our float bowl. Taking the carb apart, we discovered the float had filled with gasoline. Float turned sinker, as it were.

"There's a old boy just out of Winona has a Model A," the gas station owner told us. "Right by the road. You can't miss it. He might have parts."

We leaked on down the road and, sure enough, there was a nice 5-window coupe parked under a carport next to a house. As we pulled into the driveway, so did a tow truck with a dead Oldsmobile on the hook. The owner of the

place, Raymond George, ran a towing service and used furniture and saddle repair shop.

"I don't think I've got an old float," he told us, "but we can try to get the gas out of this one and resolder it."

As the solder sizzled, he said, "You know, I once bought 27 Model A Fords and one old Dodge for $2.50 each from a junkyard that was going out of busi-

> *After all these years, the old Ford is still a car that belongs . . . an oddly potent reminder of good things to a great many people.*

ness. I brought them back to my garage and cut them up for scrap."

"How did you move them?"

"Drove most of them, and towed the dead ones with the ones that ran. I didn't have a cutting torch, so I cut them up with an axe. When it was all over, I cleared $1.80 on the whole deal. One dollar and eighty cents."

He refused to take any payment for the repair. We said goodbye to Raymond George, and he went into the house to have the dinner his wife had held for two hours so he could help us.

We thought our mileage would improve with the float repaired, but the next fill-up computed out to 16 mpg, about what we'd been averaging earlier. People who asked us about our mileage were amazed that it was so low and would tell us "That thing should be getting 25 or 30 mpg." Not so, other Model A buffs would tell us. "They've gotten used to Honda Civics. Sixteen mpg used to be considered good mileage, back when most cars got 10 or 12."

We made it to Thayer, Missouri late that evening, just in time to get a motel room and a grain-fed catfish dinner before the cafe closed. The next day, we breakfasted at the State Line Steak House & Truck Stop and met a friendly fellow named Richard Goans, a retired postman and Model A buff, who had coffee with us. When he learned that we had a patched carb float and no crank handle in case of starter trouble, he put down his coffee cup and said, "I'll be right back."

Twenty minutes later he walked into the cafe and laid a carb float and a crank handle on the table. "No Charge," he said. "I just wish I could climb in there and go with you. At least until you get to the LA freeways. Then you're just going to have to pull your hat down tight over your eyes and hang on."

After driving south into the Arkansas Ozarks, we turned west toward Oklahoma. Our old map showed a ferry crossing at Norfork Lake, but a new bridge had replaced it. We climbed a hill just after the bridge and suddenly the engine lost power and the exhaust note went from its normal uphill chug to a strange doop-doop-doop sound. We coasted off a side road and down into the parking lot of a lovely old stone-and-timber lodge with a spectacular view of Norfork Lake. "Good spot for a breakdown," I said. "The Royal Motel. Whatever's wrong, we're staying here for the night."

We pulled the plugs and discovered two cylinders aspirating together. A blown head gasket. We checked into the motel and called an auto parts place in town. They didn't have a head gasket, the man said, but there was an old boy named Gerald Cooper, a Model A owner, standing right at the counter, and he might have one. Mr Cooper said he didn't, but we should call Bill Dean at the feed store. Bill Dean had one, and when we asked if he could send it out to the motel by cab, he said, "Cab? We don't need a cab. I'll be right out."

Fifteen minutes later Bill Dean showed up in his pickup with a new head gasket, and Gerald Cooper and his wife Ruth also pulled in. Mr Cooper had

MODEL A ODYSSEY
Part 1

Above: Moods of a town, written on the water towers. Right: Author filling tank and no doubt soaking tennis shoe.

alerted the media, and soon Kristin Turrill of the *Baxter Bulletin* arrived to do a story on our plight. Dean and the Coopers stayed all afternoon to help install the head gasket, while Ms Turrill took pictures.

That evening, Bill Dean invited us home to see his Model A collection in its new garage/storage building and his field of Model A bodies, axles, wheels, etc. The Coopers invited us back to their house to see their beautiful Model A coupe, and later Kris Turrill took us out to dinner at a place called the Back Forty. After we returned to the hotel, Chris and I sat on the porch overlooking the moonlit lake and quietly played guitars. (No easy feat, when you can barely play in broad daylight.) We agreed that next time we came to Mountain Home, Arkansas, we would sabotage our own car.

We didn't have to, however. It sabotaged itself. The next morning, on the same hill, the generator burned up. An internal wire had come loose. This time we motored straight into town and parked in Bill Dean's driveway. He found an old generator and we put one good one together from mixed parts, at Bill's Generator Shop nearby. We had lunch at the Back Forty (best hamburger of trip, "the Barnburner") and said goodbye to our friends.

As we drove out of Mountain Home, I said to Chris, "Do you remember that line in *Easy Rider* where Jack Nicholson says this used to be a good country but he doesn't know what happened?"

"Yeah?"

"Well, either he was wrong, or it's gotten better."

A day later we clattered into Okemah, Oklahoma, boyhood home of the late Woodrow Wilson Guthrie. It's a pretty little town in the east Oklahoma hills, and it has three water towers, variously marked "COLD," "HOT" and "HOME OF WOODY GUTHRIE." We looked for Woody's home, but an elderly couple who were gardening told us the house was gone. "A local businessman who didn't think much of Woody bought the property and tore the house down before the town could stop him," they said. "Folks here either seem to love Woody, or they don't care much for him." We admired their garden, and the couple gave us a jar of homemade strawberry jam to take with us.

"Present tense," I said as we drove away. "They talk as though he's still alive."

"Maybe he is."

"Well, he'll certainly be better remembered than the brilliant guy who tore his house down."

"Whose name escapes me."

At a main street gift shop in Okemah, I bought a hat that said, "Okemah, Oklahoma, Home of Woody Guthrie," with a picture of the water towers. "Arlo Guthrie came through here about a year ago," the shop owner told us. "He'd seen these hats at a concert and wanted a couple of them, so we opened the store up on a Sunday so he could buy some. He seemed like an awfully nice fellow."

As the most malevolent-looking wrath-of-God thunderstorm I'd ever seen gathered over town, throwing bolts of lightning at nearby hills, we drove south toward Texas. Chris got out his guitar and we tried to remember all the verses to "This Land is Your Land." My favorite lines of that song, in the verse so often edited out for children and nervous adults, went:

"And on the sign it said
'No Trespassing'
But the other side didn't say nothing
That side was made for you and me."

Fun, sly humor. The kind that got your house torn down.

A fierce wind shook the Model A, cuffing its slab-sided bodywork all over the road and making steering twitchy. The car suddenly felt very tall as it pressed on through the wind and the lightning; a mobile lightning rod shaped like a billboard. Old boxes and uprooted bushes blew across the road too fast to tumble. Chris put the guitar away, chords drowned out by wind and gusting rain. "Maybe you'll get that tornado you wanted."

"Somebody's getting one," I said, watching the sky go dark as night in the rearview mirror. The three water towers on the hill stood out eerily, bottle green against the blackness. A jagged bolt of lightning cracked down just behind the town. "Looks like that's the end of Oklahoma," I said, advancing the spark. "Let's go to Texas." ◆

Next month, Part II: Halts across Texas, diamond deserts, lonesome highways and Knockin' on Heaven's Door.

So long, Okemah, it's been good to know you . . .

MODEL A ODYSSEY

Part 2

Halts across Texas, high times in Luckenbach, lonesome highways, diamond deserts and A flats

BY PETER EGAN
PHOTOS BY THE AUTHOR & CHRIS BEEBE

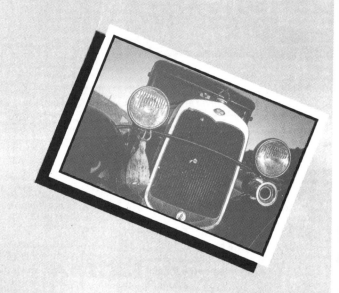

In our last installment, a 1930 model A Ford loaded down with Peter Egan, Chris Beebe, two guitars and an inadequate supply of spare parts was chuffing south through Oklahoma toward Texas on a 3300-mile jaunt from Wisconsin to California. We find our two drivers now driving through a tornado-ridden afternoon in Dust Bowl country, surviving only on their wits and a small handful of credit cards.

WE DIDN'T SEE any tornados that afternoon. It was too dark. But if we'd had a radio in the Model A, we would have learned that a tornado was flattening the town of Saragosa, Texas at that very moment, a few hundred miles to the southwest. We sawed at the wheel in high winds and skittered south through Oklahoma until the sky cleared and rainbows appeared over Highway 78, right at the border. "Red River," the sign read, and on the other side of the bridge it said, "Welcome to Texas."

Ah, Texas.

When we told people we planned to drive a Model A across Texas, a common reaction was, "That's too bad. It's big and flat and dull and hot. You'll go crazy with boredom."

The only part of Texas I'd ever seen was that short stretch of Interstate 40 that cuts through the northern Panhandle, so I couldn't vouch for the Pan itself. But I figured any state that produced Willie Nelson, Waylon Jennings, Buddy Holly, Lightnin' Hopkins, Doug Sahm, Blind Lemon Jefferson, Bob Wills, Kris Kristofferson, T-Bone Walker, Jerry Jeff Walker, Johnny Winter, Janis Joplin, Stevie Ray Vaughan, ZZ Top, Joe Ely, Austin City Limits, the Terlingua Chili Cookoff, Jim Hall, Carroll Shelby, A.J. Foyt, Hap Sharp, Cobras, Chaparrals, The Fabulous Thunderbirds, Sam Rayburn, Larry McMurtry and Lone Star beer couldn't be all bad. About half of the music I listened to came out of Texas (the other half coming from the Mississippi Delta and the South Side of Chicago). The place had to have something going for it.

However favorably disposed I might be toward Texas, there was no doubt that it was a big place to tackle in an old car. It filled two full pages of our road atlas. "If the Model A were the size of an Isetta," I explained to Chris after studying the map, "El Paso would be over a billion light-years away."

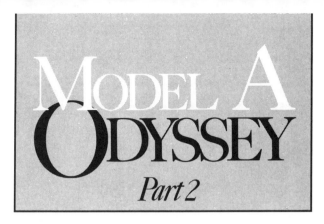

MODEL A ODYSSEY
Part 2

We drove south through green, rich farm country that seemed more Midwestern than Western, passing anti-littering signs that said, "Don't Mess with Texas." None of this, "Thank you very much for not littering please and have a nice day" stuff you got in other parts of the country.

Our first stop for the night was in Bonham. After checking in to a motel we filled up at a corner gas station and asked if we could use their jack to check the front end of the Model A, as it had been developing a shimmy. The owner not only lent us a jack, but also helped us find a loose bolt in the front A frame and lent us a grease gun. "I've had some good times in these cars," he said, gazing wistfully into the interior. "Also a couple of bad wrecks. A bunch of us crashed into a grove of trees on our way to a wedding and couldn't open the doors. Had to cut through the top."

Off toward Austin the next day, skirting the suburban sprawl of Dallas, we passed a city-limits sign that said, "Welcome to Avalon: Home of the Filthy Rich," and traveled southward on roads scattered with dead armadillos. We were told they come out at night and jump straight up when a car passes over, so it does no good to straddle them with your tires. With the ride height of the Model A, I suggested they might have to pole-vault to kill themselves. "I wonder if we'll ever see a live one?" Chris asked.

"Not with these headlights," I said. The Model A beams were not very powerful. They were the only headlights I'd ever seen where you had to get out of the car at night and check them with a flashlight to see if they were working.

Rolling into Austin in late afternoon, we found ourselves in the University area, lunching at a student coffee house called "Dr. Quackenbush." Chris called an old friend named Marti Dapin, who joined us for coffee and then showed us around the charming city of Austin. We had a chicken-fried steak dinner at the famous Threadgill's (best meal of trip), a famous Western-style bar and restaurant where Janis Joplin, among others, got her start. We then checked in to the Driskell Hotel (best hotel of trip), a beautifully restored place built by a cattle baron in 1886, and that night we wandered down Sixth Street, which has more bars and live music per square foot than any place I've been.

Lone-Starring our way through the evening, we learned that Marti is a lighting technician on the *Austin City Limits* program, and in the morning she gave us a tour of the sound stage where it's taped. Chris and I stood on the very stage, thinking it would be nice to tell our grandchildren about if we ever had any and we could get them to stop listening to Madonna's granddaughter long enough to pay attention.

After lunch at a great little bakery called Texas French Bread (the state name is used frequently in conjunction with business names to add impact, even more than in California), we said goodbye to Marti and headed straight west.

West of Austin you run into the famous Hill Country, an area of scenic ranch land full of small streams, rock outcroppings, woods and green valleys. We drove out to Johnson City, a small town that was the boyhood home of LBJ. His modest but finely built family house is still there, and I sat for a while on the front porch swing and rocked, looking out at the pleas-

ant yard and enjoying the cool shade. When we left, I told Chris, "Well, I finally got something out of LBJ, other than a draft notice."

"Fair trade?"

"Nope."

We cruised past the LBJ Ranch on Highway 290 and stopped for gas. I asked the attendant if he knew the whereabouts of Luckenbach, Texas.

"Souvenir hunters have stolen all the signs," he said, and then gave us an elaborate set of backroad instructions to find the place.

For those who don't follow that kind of thing, Luckenbach, Texas is a little one-horse town made famous in a song of the same name by Waylon Jennings. The town was bought in the Sixties by a Texas rancher, humorist and all-around character named Hondo Crouch, who thought it would be fun to own his own town, post office and all. He promoted concerts there and turned it into a sort of Woodstock of country music.

Chris and I made a few wrong turns, but finally found the place, a small collection of weathered board buildings—bar,

> ***They were the only headlights I'd ever seen where you had to get out of the car at night and check them with a flashlight to see if they were working.***

post office, dance hall, general store and a few houses—in a wooded glen near a stream. There were a lot of pickup trucks parked outside the bar. We parked the Model A and went in.

Most of the people were out in back, in a kind of beer garden under the trees, a relaxed Sunday afternoon mixture of local ranchers, sitting around picking guitars, and a group of bikers who were there with their families. Chris and I had a beer and listened awhile, and then Chris said to me, "Let's go get our guitars out of the car."

We spent the rest of the afternoon and evening playing guitars with what turned out to be a remarkably nice bunch of people. There was a woman named Julie, who had a voice like Judy Collins; James Luckenbach and Jim, Jr, a father and son who owned a nearby ranch and knew the words to more Hank Williams songs than the Library of Congress; and a biker called "Coyote" who bought us several beers. When the place closed, a rancher named Ken Morgan came up and asked where we were staying.

"I guess we hadn't thought about it."

"Well, if you've got a tent, you can camp at my ranch. The house is all torn up right now for remodeling, but you can camp at a little lake I've dammed up near the house. Just follow my pickup in your Model A."

We set up our tent by Ken Morgan's pond and sat up drinking with him at the ranch house until the wee hours, learning that, in addition to being a cattle rancher, he was a retired airline captain and was in the midst of restoring a Jaguar XK-150 and that Ken and his wife, Kathy, presently owned half of the town of Luckenbach and the daughter of the late Hondo Crouch owned the other half. His nephew David, who was spending the summer at the ranch, was an avid R&T reader, and David's mom in Dallas was having Campagnolo build a special set of wheels for her new Testarossa. Just your average Texas ranch family.

Retiring to the tent, we discovered we'd camped right next to

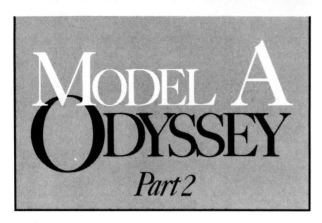

MODEL A ODYSSEY
Part 2

West Texas oil country on Highway 33; our 16-mpg Ford rumbles past its future.

a sort of armadillo freeway at the edge of the pond, with lots of the little critters caught in our flashlight beam as they rustled by in the tall grass. "The best thing about armadillos," I observed, "is that they aren't rattlesnakes." In the morning we went for a swim in the pond, and then Ken took us out to breakfast at a small cafe in Blanco.

A few hours later, Chris and I stopped in Mason, Texas at a hat shop and got ourselves a couple of hats, steamed to shape. After three days, we'd discovered we were the only males above the age of three in rural Texas who were not wearing white hats at all times. As a when-in-Rome cultural observation, it should also be noted that adult men in Texas ranch country rarely wear polo shirts with small animals on the pockets, short pants, sunglasses hanging from straps around their necks or running shoes, except in the case of Willie Nelson, who can get away with it.

Suitably hatted, we duded our way west, with plenty of head room left in the Model A. We followed Highway 190 through the oil country of the Permian Basin and then turned north on Highway 137 toward Big Lake. We had a flat tire, bolted on one of our two spares, and rolled on.

Chris stopped in the middle of nowhere to take some pictures and left the car idling, while I leaned against the front fender. Listening to the engine, I heard a curious new ticking sound and opened the hood. As I stood gazing at the engine, the water-pump pulley exploded with a sound like a gunshot and the fan threw off a chunk of metal that glanced off my chest and into a nearby field. (Thereby reversing the usual metaphor about fans and fertilizer.)

I was undamaged, but the pulley was history. I tried to restart the car, but the starter picked that moment to die of a broken drive spring. So we bump-started the Model A and chugged into a Fina station in Big Lake, Texas with the remains of the pulley chewing up the remains of our belt.

The underground Model A good-old-boy network came through again. Station owner Tom Cantrell called up Ford buff W.E. Short, who didn't have a pulley but called Red Nutter at his metal salvage yard. Red Nutter pulled up at the Fina station in his pickup five minutes later and said, "Hop in. I know an old boy out in the oil fields who has the parts you need." We rode 30 miles out in the country and met Bill Miller, who has a Model A and a backyard full of Ford parts.

"Watch out for the rattlebugs," he said as we searched the backyard for a water pump pulley.

"Rattlebugs?"

"Rattlesnakes," he said, reaching into an old brown paper bag and pulling out a Model A water-pump pulley. "The yard's full of 'em."

Chris and I retreated from the yard with our eyeballs gimballing around like independent searchlights.

Tom Cantrell made us keep the Model A in his garage all night, free of charge, so kids wouldn't mess with it, and the next morning he, Short, Nutter and two guys who worked at the station, Mike Hunter and Tony McEntire, helped us fix the pulley, change oil, adjust valves, repair the starter and fix our spare. The garage charged us $30 for oil, fuel, parts and labor and we were on our way.

Climbing the Caprock Escarpment, a wall of rock that separates the Permian Basin from the high plains or Llano Estacado, we drove into Lubbock in the evening and burned out a headlight trying to find Stubb's famous rib place, which had changed hands and is now full of college students watching MTV on a big screen. In the morning we cruised past the statue of Lubbock native Buddy Holly on our way to a guitar shop. Since Holly and Waylon Jennings both bought their first guitars here, we thought it would be a good place to look around for a good Strat or Telecaster—historical significance and all that. The music store where we stopped carried nothing but Korean reproductions of Japanese guitars that were copies of American designs, so we motored on into New Mexico and across the Pecos River.

"No law west of the Pecos," I muttered reflexively as we crossed the bridge. "Whoever said that never tried speeding on the Ventura Freeway."

After driving through the pretty town of Roswell, we turned off Highway 70 and climbed into the mountains to the little town of Lincoln, New Mexico. Lincoln is an authentic, very original old Western town, famous as the center of the Lincoln County cattle wars and the place where Billy the Kid shot his way to freedom and escaped from the Lincoln County courthouse. The graves of the two men he shot, Bell and Olinger, are still right next to the courthouse steps. Dozens of movies have been made about this episode, the best, in my opinion, being Sam Peckinpah's *Pat Garrett and Billy the Kid*, which I've seen about 15 times, so I could be biased.

Chris and I checked in to the lovely old Wortley Hotel, had a steak dinner and spent the rest of the evening sitting on the upper porch of the Lincoln County courthouse, playing guitars. It was still too early for the tourist season, and the place felt almost like a ghost town, with the chords of "Knockin' on Heaven's Door" echoing off the empty, ancient wood walls of the main street. Dogs howled pitifully, but there were no people to complain.

It was warm the next afternoon when we dropped back down to the plains, so I got out and rode on the running board for a while. Chris set the throttle lever and also got out, steering through the window. Then we slowed the car down to a walk and switched sides.

When running boards disappeared from cars, I concluded, we lost a wonderful institution. On this trip alone, we'd found dozens of uses: a footrest for people who gave us directions, a way to climb aboard for short hops from gas station to cafe, a place to set luggage out of the puddles while loading the car in the rain, a workbench, a parts shelf, a bench for sitting in the shade while reading a map, a footrest for tying your shoe, a platform for shooting pictures and a place to set a drink, smoke

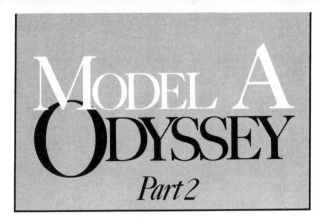

MODEL A ODYSSEY
Part 2

a cigarette, play a guitar, look at the scenery or make a quick getaway from a bank robbery. Running boards, dollar for dollar, added more fun value to a car than all the handling packages, sticky tires and sound systems in the world.

Running boards also worked fine for laying out carb parts, as we discovered, just past Quemado, New Mexico. We were climbing a grade outside of town when the engine made a loud rattle and the throttle suddenly stuck open. Dissection of the carburetor revealed that the engine had swallowed a throttle butterfly screw, leaving the butterfly crooked and stuck. We did a roadside repair with wire and a flat rock and drove back to Quemado. The owner of Allison's Motel called a rancher named Jim Williams who drove into town, got out of his pick-up truck and handed us a complete Model A carburetor. We'd already found another screw at a gas station and fixed the problem, but Jim said, "Well, take the spare one with you and mail it back to me if you don't need it." He waved goodbye and said, "I got to go saddle up a horse and check my country. Some of my cattle got through a fence somewhere."

As he drove away, Chris said, "I can think of worse ways to spend a day."

Model A fixed, we spent the evening in Ol' Jonay's Quemado Tavern, where I taught Chris how to play pool, while trying to remember myself. We played the world's longest game of pool while cobwebs grew in the side pockets and the local patrons urged us on with drinks and free advice, and Dwight Yoakam and Steve Earle sang on the jukebox. It was an evening to remember, what little of it we could later recall.

Into Arizona, through Show Low and down into the Fort Apache and San Carlos Indian Reservations on dirt and gravel roads. Beautiful mountain country, full of small ranches, pine trees, mountain streams and men on horseback, land of the two great Apache chiefs, Cochise and Geronimo, who fought ferociously to hang on to their homeland. Then down out of the mountains and onto Interstate 10 into the mercilessly endless sprawl of Phoenix. I looked out at the thick smog, trailer courts, manufactured housing, fast food franchises, carpet warehouses, satellite dishes, waterbed emporiums and suburban clutter of the city and said to Chris, "Tell me again; did the U.S. Cavalry win the Indian wars, or was it the Apaches?"

We stayed at a chain motel on the interstate that night, just west of Phoenix, and made it all the way to Newport Beach the next day. Not much to say about the I-road, except that we made it though the California border station with a lot less hassle than the Okies (we had that do-re-me), and the Model A climbed the grade at Desert Center without overheating or blowing up. Also, while we were gone somebody had changed all the speed-limit signs from 55 to 65. And us in a 45-mph Model A. We got to the Coast in the early evening, clattering home in rush-hour traffic. We gave my wife, Barbara, a ride in our trusty car, then went out for Mexican food.

In 3370 miles the Model A averaged 16 mpg, cruised at 45 mph, used four quarts of oil (plus an oil change), had two flat tires—both caused by the chafing of a sandblasted, formerly rusty rim on the inner tubes—and six mechanical failures. We had a sunken carb float, blown head gasket, burned generator, broken starter spring, exploded water-pump pulley and a swal-lowed throttle butterfly screw. Oddly, all of the mechanical problems were related to the only aftermarket non-Ford parts on the car, pieces that were either substandard or improperly assembled. The carburetor, for instance, was an aluminum "updated" unit, installed by a previous owner who thought it was more efficient than the old cast-iron Ford carb. Same for the potmetal water-pump pulley, a lightweight replacement for the Ford steel version of same. The off-the-shelf rebuilt generator had a screw installed wrong, so it bottomed on the generator casing, and so on. With a completely original Model A, we'd have had exactly zero problems.

Good thing we didn't, though. It was the breakdowns that allowed us to meet people in Mammoth Spring and Mountain Home, Arkansas, in Big Lake, Texas, and in Quemado, New Mexico. It was the breakdowns that made the trip, that got us invited into homes and backyards and garages by people we'd never have met if we were driving something more sane and modern. Also, every problem was fixable, with a few simple tools and a few used parts that someone had lying around on a shelf or in a backyard. The head gasket was the only new part we used.

> ## "Tell me again; did the U.S. Cavalry win the Indian wars, or was it the Apaches?"

Richard Bach once said that old airplanes live forever, unless crashed, and have a much longer lifespan than human beings. The good ones—the simple, honest designs—can be rebuilt over and over again.

The Model A, I think, is like those old planes. Given reasonable care, it should outlive all of us, and if the supply of gasoline holds out, it could be chugging across the country again 58 years from now. Which is more than you can say for most of the cars that passed us on the highway. When Henry Ford built the Model A, he ignored his accountants' pleas to use lower-grade materials and make the car cheaper. He told them that if you made a high-quality product, the public would recognize the quality, buy more cars, and the profits would take care of themselves. Ford was a visionary who wanted to build something permanent, a machine with lasting worth and value.

After all these years, the vision is still a good one and the Model A remains a fine car. Like a Woody Guthrie song or a good guitar, it doesn't get old, it just gets better. ⊚

46e GRAND PRIX DE

MONACO

Monaco for beginners

BY PETER EGAN
PHOTOS BY THE AUTHOR

"**B**EFORE I DIE," my friend George Allez told me a few months ago, "I'd like to go to Monaco, just once, and see the Grand Prix."

Fortunately George is in excellent health, so his wish to see Monaco can probably be interpreted as a long-term plan rather than a deathbed request, but I know what he means. If you grew up on Formula 1 racing, Monaco by any other name is Mecca. I've wanted to go there myself ever since I saw Stirling Moss and Richie Ginther battling across the pages of R&T in 1961, and until last week I'd never been there.

That's not quite true.

My wife Barbara and I tried to alight in Monaco about 15 years ago. It was on our slightly delayed 1973 honeymoon tour of Europe, when we drove 8000 miles in a rusted-out, over-heating $250 hulk of a Simca 1000, bought in Amsterdam from a man who is now burning in Hell if there is a just God.

Divine retribution aside, we sputtered into Monaco with this delightful car, trailing coolant and smoke, and soon learned that we would not be allowed to park. Anywhere. Every time we pulled the Simca over to a curb, a policeman or a bell captain or some other uniform-resplendent figure walked up to our car and shook a white-gloved finger at us in a sort of metronome motion, that uniquely European gesture of denial. The authorities simply took one look at our rusty Simca, our shining Youth Culture faces and Dutch plates and sent us right on down the road. After an hour of being turned away from parking spots, we gave Monaco a healthy Bronx cheer and removed ourselves to Italy, which proved to be as friendly as Monaco was hostile.

Monaco's race fans have always had a deep, abiding love affair with the sea.

But I never bear grudges, especially where racing is involved, so when Editor John Dinkel offered me a chance to go to the Grand Prix this year, all was forgiven. I ordered the plane tickets, got reservations at Loews Hotel and asked Barb to come along. "This time," I assured her, waving our hotel reservations, "they've got to let us park."

At LAX we boarded a British Airways 747 for Nice, via London, Steerage Class, and I found myself pinned in my seat by the recliner in front of me like something in a bench vise. (Whatever corporate bottom-liner sneaked in those extra 20 or 30 rows of seats should try a ride in one of his own airplanes.) Nice stewards and stewardesses, though, so I had the feeling of being visited by cheerful friends while laid up in the hospital with a full body cast. At London we transferred with some relief to Air France, dozed and awoke on final for Nice, the nearest flat ground to Monaco, inbound low over the impossibly blue Mediterranean.

After LAX and Heathrow, Nice is a pleasantly small airport that reminds you of the last scene in *Casablanca*. The car-rental people handed us the keys to a Fiat Panda, which took a deep breath and hauled us up a mountain to the ancient cliff-clinging village of Eze. There we splurged on a 4-star hotel and restaurant called the Chevre d'Or, figuring we could sell our house to pay the Visa bill when we got home. The next day we descended from the mountaintop into Monaco, our Panda geared down and backfiring softly, like Mr Hulot's Amilcar.

Cruising in heavy traffic and working our way downward to the sea, we found ourselves between two rows of Armco with the curbs painted red and white. "This is it!" I said to Barb. "We're on the circuit! This is the Monaco tunnel, where Innes crashed . . . here's where Ascari went into the sea . . . the old Gasworks Hairpin!" and so on. Barb patted me on the knee, a gesture perfected over years of enduring this sort of adolescent marvel. We finally found Loews Hotel, a big glittering place whose north side looks right out on what used to be the Station Hairpin, when there was a train station. Now the corner is called Loews Hairpin. Just below the hairpin, the circuit snakes down to the waterfront and enters the tunnel, which passes directly under the hotel.

We checked in, parked our car (yes!) in a public garage next to the tunnel, hooked up with Innes Ireland, who had also just come into town, and headed out to walk the circuit.

Even a week before the race, you could feel a sense of deadline fever in Monaco, with workmen bolting Armco and grandstands together while the police directed traffic around cranes lifting huge concrete barriers into position. Everywhere people walked the course, looking forward and backward, comparing possible viewing spots for the race and doing double takes when they spotted Innes.

A strange, fascinating city, Monaco. Palm trees, coral roofs, lovely old Belle Epoque architecture with balconies, domes, pillars, porches, awnings, parks and promenades that Henry James might have used to make Daisy Miller feel at home, all mixed in with hideous poured concrete modern hotels and apartments of the East German Worker Housing variety. It is clear that at some point the principality lost its nerve and abdicated all hope of controlling its architectural destiny, possibly for money. Happily, most of the brand-new construction in town seems to show a return to the charming, ornate style that once made Monaco distinct from, say, postwar Leningrad.

The harbor, of course, was full of the yachts of the rich and/or famous, the largest being a rakish white craft about the size of the *Graf Spee*, a ship that Innes said belongs to Donald Trump, by way of the oil sheik Kassoghi and the Sultan of Brunei. "I wonder if I should have insisted on something larger than a Fiat Panda?" I said to Barb as I focused my camera on the ship. "But what the hell, it got us here, same as him."

We walked along the harbor portion of the circuit, slowing for the chicane, and then down to Tobacconist, a famous turn that still has a little tobacco shop built into the terraced masonry. "The circuit used to come right past this shop and continue along the waterfront beneath those grandstands," Innes said, "before they put in that great bloody pool and slowed everything down."

The tobacco shop is now well back from the circuit, hidden behind a wall of Armco and tall chain-link fence. Not as charming as in the old days, but not as ugly and sealed-off as many modern street circuits. Monaco has elevation on its side. If you can't see through the fence or over the barriers, you simply seek higher ground, watching from a terrace or a balcony. Or the well-placed grandstands, for which you pay about $180 for your place in the sun. Or rain. With a little altitude, the fences disappear from view, and you can almost imagine again that the point of racing is not to make everyone extremely safe.

As we rounded the old Gasworks Hairpin, which is now called Rascasse (in honor of the new Restaurant Rascasse and the missing gasworks), Innes recalled his various shunts, brushes with stone and other Moments at Monaco. The worst, of course, was his 1961 selection of 2nd gear when he wanted 4th in the Monaco tunnel. His Lotus emerged from the tunnel in a sort of shotgun blast of stray parts, leaving Innes back in the darkness to cope with traffic and two broken knees. Another time, he got out and pushed his car nearly a full lap after the ignition failed.

Fireman James Earl Jones (top) looks on, but the Force was not with Mansell (bottom), who lost 3rd to Alboreto.

"Christ," Innes said as we climbed toward the Casino, "I can't believe I pushed a car all the way up this bloody hill. It's all I can do to walk it now, without the car." It was all I could do, too, so we stopped in Rosie's, a famous old trackside pub, for a drink.

That night we had dinner at the hotel and then walked up to the Casino for a look around. The front door featured many pictures with red slashes through them. I stopped to see if they'd let us in and found no prohibition against our entering, as we did not have a dog, *were* wearing shirts and shoes, but not T-shirts, running shorts or backpacks, and had no plan to videotape our gambling experience. There was no picture with a red slash through a journalist with *sauce provençale* on his necktie, so the doorman let us in.

The Casino has sort of a dowdy old elegance mixed with Bally slot-machine glitz, though it gets classier in the back rooms, where the green-felt table games are played. I contributed a few 5-franc coins to the slots, but avoided the tables. I steer clear of playing expensive games with anyone craftier than I am. This is a sizable group of humans that, so far, includes everyone I've ever met.

Thursday morning, the first day of practice:

Cars on the track at last! I met Innes in the lobby and we walked out into a dull drizzling rain to watch practice from Loews Hairpin.

Nelson Piquet's Camel-yellow Lotus appeared first, throwing spray off its tires in the Mirabeau, then came the sweatshirt pastels of the Benettons, the electric-orange Marlboro McLarens of Ayrton Senna and Alain Prost, the Ferraris of Gerhard Berger and Michele Alboreto, still in pure national Italian red. They came twitching and feinting down the rain-slick chute to the hairpin in alternate bursts of wheelspin and slewing oversteer, like a bunch of high-strung 600-horsepower pigs on ice. Angry whoops of engine noise bounced off the stone walls with each lunge of acceleration. I didn't envy the drivers (not in this weather, at least) trying to keep the cars on the road. It was chilling to watch.

It took 5 or 6 laps in the wet before most cars wore the white stickers off their tires, but then the track began to dry out and the pace picked up. "Boy," I said to Innes, "Mansell looks fast." Then I said, "Wow, Berger looks fast. And Palmer's really pushing, too. So's Warwick." A few minutes later, I said, "Hell, they all look fast."

When the first practice session was over, Innes and I hitched a ride to the paddock, where I discovered that hospitality tents now occupy more square footage than the transport trucks. Innes and I had a sandwich in the Goodyear motor home, then walked up to pit row for the second practice session. We stopped to watch half a dozen mechanics doing a last-minute adjustment on Senna's engine, probing its innards with a small T-tool and blipping the throttle in earsplitting barks. Another six or seven McLaren mechanics worked on the car itself. "Look at this," Innes said. "When I came here with Lotus in the early Sixties, we were lucky to have three mechanics for two cars."

Whatever adjustments the mechanics made must have been the right ones. Senna went out in the dry that afternoon and drove about 2 seconds a lap faster than any of the other drivers. He turned a 1-minute 26.46-seconds lap, and his teammate Prost was second fastest with a 1:28.37, just a little quicker than Mansell, who appeared to be driving the wheels off his Williams.

That evening we had dinner at a very nice little side-street restaurant called L'Etoile, and then walked up the circuit to another famous watering hole called the Tip-Top, which is located on the downhill straight between the Casino and the Hotel Mirabeau. The Tip-Top is a small bar where all the F1 lads hang out after work, and when we got there, the crowd was swelling into the street. There was a raucous party in progress, honoring two Team Lotus tire changers named Clive and Kenny, who had been with the team for 10 years. Clive & Kenny showed up in Benny Hill-quality drag—wigs, high heels, sunglasses and the whole bit—posing as the two ugliest women in the world. Several people tried to take liberties and were clouted with handbags.

We moseyed on over to Rosie's where Innes threw a fit because they tried to serve him a beer in a plastic cup. "I'd like a proper glass, please," he said. The manager explained that they had to use plastic on race week because people might throw their glasses onto the track.

"I am a former Grand Prix *pilot*," Innes explained patiently, "and I don't throw my beer glass onto the bloody track. You can trust me."

The manager shrugged helplessly, so Innes shoved his beer can and plastic glass back across the bar and said, "Let's go. This used to be a quality place." **Continued on page 91**

Continued on page 91

AU REVOIR VITESSE

Can a Citroën 2CV find happiness in the Canadian Rockies? Almost indefinitely.

BY PETER EGAN

PHOTOS BY THE AUTHOR & CHRIS BEEBE

"SAY, THAT'S QUITE a rig," the gas station attendant said. "You boys make that little truck yourself?"

"No sir. It's French."

"Ah."

We were, at that moment, in the middle of North Dakota, but we'd been accused of personally constructing our 1956 Citroën 2CV van by people who lived in such diverse places as Seattle and Burnt Flat, British Columbia. Apparently no one wanted to believe that an honest-to-God professional car company would bother to gear up its forges, stamp presses and assembly lines to produce a machine that looked, essentially, like a tree sloth trying to escape from a Quonset hut.

Until we told them it was French. That seemed to make it all right. In a country where children drank wine and Jerry Lewis was regarded as a comedic genius, anything was possible. Even a car called Two Horses.

We paid for our gas, calculated the mileage—36 mpg this time—climbed back into the car and slammed the suicide doors. The fold-up window swung down like a trap and tried to smash my fingers, but I was too quick for it. A full week of crunched knuckles had made me wary as a cockroach at a flamenco contest.

A full week. And we were only halfway home.

As usual, this whole clambake was the idea of my old-friend-and-former-employer, Chris Beebe, from Madison, Wisconsin. Chris is a foreign-car repair-shop owner, gentleman farmer, musician, connoisseur of silos and collector of odd cars and busted riding lawnmowers. He'd called me a few weeks earlier to say that his brother and sister-in-law, Russell and Trinka, were selling their 2CV truckette. Chris was flying out to Medford, Oregon to get the car and drive it home. Did I want to come along?

"Well, I've never driven a Deux Chevaux," I admitted. "On

Above: Trinka, Russell and daughter Mi-Lan say goodbye to a friend. Right: Teepee and sweat-lodge-in-progress.

the other hand, my last two cross-country trips have been in a Model A Ford and a Piper Cub. If I take one more ridiculously slow trip across the country in a living antique, I may suffer some kind of permanent brain damage. If I haven't already. I feel we're tampering with Relativity here. Doesn't your brother own anything faster than a Deux Chevaux?"

"No. He owns three of them," Chris said. "But this one's pretty fast. It's got the later 600-cc engine in it, instead of the original 300-cc engine. Russell says it can go virtually 65 mph. On level road."

"Gee. That fast. Okay, I'm hooked. Let me talk it over with the Editor."

A week later I was flying north along the Sierras, which were on fire. It was autumn, after the driest summer in decades, and smoke from dozens of forest fires made the sun shine with an unearthly copper glint all the way up the West Coast. Descending past Mount Shasta, the plane landed in the lovely valley of orchards and farms where Medford is kept, the air heavy with cedar incense.

Chris's brother, Russell, met me at the airport, pulling up in an immaculate green 2CV sedan with a beautiful handbuilt wooden dash and door panels. Russell designs and builds fine furniture for a living, and, like Chris, his eye for balance and detail touches everything he owns. It's a trait that runs in the family. Their parents are both artists, so the whole clan can draw, paint, sculpt, etc. Impressive stuff, when you can barely put a model airplane together without getting half the parts stuck to your own forehead.

"Chris couldn't come along," Russ explained. "He's been at the garage of my friend Mike Warne all week, working on the 2CV. The car's been sitting for a long time."

When we got to Warne's garage, Chris looked hot and frayed, as if he'd been rolled down a hillside in a footlocker full of old transmission parts. He apologized for his T-shirt, which was throwing off death rays.

"Two more days, and this thing should be ready to go," he said. "I've already installed a new clutch, axle joints, brakes and a starter. I'm still working on the rear-axle bearings, which are almost impossible to do without a special Citroën tool."

"Do we have the tool?"

"No."

I walked into the garage and looked at the Citroën, which was wheel-less on jackstands. I'd seen this neat little truck once before, in 1975, when Russ and Trinka showed up in Madison at the end of a 2-year trek that had taken them to the bottom of South America and back. It was painted a caramel-beige color, had a spare tire mounted on the hood and an interior like a Buddhist temple, full of carved mahogany, beaded curtains, candles, incense and Oriental rugs. Peering cautiously into the back, you half expected to find Bodhidharma meditating or the goddess Shiva seated on a bed of lotus blossoms. It was an idealized vision of the hippie van, with undertones of Sahara chic. It was also the kind of vehicle border police loved to disassemble, while probing its hidden corners with ice picks and trained dogs.

Chris came up and stood next to me, wiping his hands on a rag. "People were all nice to us on the Model A trip," he said. "I wonder how they'll react to a 2CV with beads and candleholders. It could be . . . unhealthy."

"Hard to say. I just hope some lumberjack doesn't cut it in half with a chain saw."

We spent the next two days replacing axle bearings, which were impossible to do without a Citroën tool, but we did it anyway, thanks to the kindness and tools of our garage-host, Mike Warne, a charming English ex-patriot who used to build Lotuses for Colin Chapman back in the Fifties. The 2CV was quite different in design philosophy from the British cars I was accustomed to working on. British cars had bearings that were easy to install but didn't last very long (10, 15 minutes maybe). The Citroën had huge bearings that looked as if they'd last forever, but installation was diabolically difficult. Tuneup and engine maintenance were easy, however, with those two simple BMW motorcycle-style jugs sticking out like yardarms.

While working on the car, we stayed with Russ and Trinka, who had an actual full-size Indian teepee in their backyard. I allowed as how I'd always wanted to camp in a real teepee, so Russ, Chris and I slept there one chilly night. Russ built a warm fire in the fire ring, and we sat around toasting marshmallows and discussing old British sports cars, much as the Plains Indians used to do.

We staggered out in the morning with matching headaches, either from marshmallow overdose or smoke. We had coffee in Russell's spacious woodworking shop, which was partly filled with the imposing hulk of a 1937 Rolls-Royce. Russ had been hired by a restorer to replace the dry-rotted ash framework inside the body. The woodwork he was doing looked too nice to cover up, and I wondered how the owner would feel about a Rolls-Royce Woody.

Chris and I loaded our guitars and luggage into the back of the 2CV and said goodbye to Russ and Trinka and their two children. "I think you'll make it without any problem," Russell said. "This car is a great trundler, and I always have faith in cars that trundle." We rolled out of the driveway and headed northwest, the plan being to visit friends in Seattle and loop through the mountains in southern British Columbia.

As soon as we hit the road, I was amazed at what a comfortable car the 2CV turned out to be. Chris said he thought the

French built the 2CV first, and then built the pavement to suit it, so they could save money on road construction. Actually, any road seems to suit it. The Citroën didn't just react to bumps; it smothered and absorbed the whole road with a soft rocking and plush jouncing motion, almost sleepily, like a mechanical lullaby. We nodded down the road in complete isolation from shock. Even the seats, essentially nothing but steel-frame lawn chairs with aftermarket vinyl covers over the original canvas webbing, were remarkably comfortable.

"What's the correct technical description of this suspension system?" I asked Chris.

"Tusk/lever/cable/push/pull/canister o'springs," he said, "with fluid globes."

Steering on the 2CV is highly castered; turn the steering wheel hard left or right as you drive down the road, and it snaps back to center immediately. Russell told us that he read *War and Peace* or some such fat novel while driving this car across the Great Plains, and I believe him. The front tires seemed to follow the camber of the road instinctively. Power is adequate, if not abundant, and the engine sound is Early BMW motorbike, with a hint of Briggs & Stratton. Shifts are thrown in push/pull motion with an 8-ball knob attached to a rod that comes at you through the dash ("Four-in-the-air," Chris called it). Shifting from 2nd to 3rd feels just like putting a buck's worth of quarters in the Maytag to start your wash at the laundromat.

Our worries about possible hippie backlash in the 2CV proved unfounded during the first two days of driving. Wherever we stopped, people thought it was a "neat little rig." Truck drivers, loggers, tourists, park rangers and waitresses all seemed to be charmed by the Citroën, and most of the many motorists who passed on the highway gave us a thumbs-up. Even a pair of camouflage-clad survivalists who drove up in a camouflaged International Scout inspected the truckette with a certain narrow-eyed admiration and asked questions about fuel mileage and traction, as though the 2CV might fit into some kind of Alternate Plan they hadn't yet considered.

Left and below: the best of all possible worlds—Eastern cabinetry, French instrumentation and Canadian scenery.

"*I think you'll make it without any problem," Russell said. "This car is a great trundler, and I always have faith in cars that trundle.*"

When they left, Chris said, "Those guys were so heavily camouflaged I didn't even see them, did you?"

"No. I just heard voices."

Circling a clear and unclouded Mount Rainier, we wafted on toward Seattle, to visit our old friends Lyman and Kathy Lyons. In the mountains, the 2CV proved a little slow on the uphill grind (35–40 mph), but made terrific time going downhill, mostly because of its benign handling characteristics. The basic handling trait is heavy understeer, but in a hard corner the 2CV almost defies physics, converting speed into suspension sway. It's a corner-absorptive car that takes g-forces and turns them to mush. You can throw it into curves at speeds that should put you right off the road, yet exit in a benign arc of soggy tire-scuff. It's the most forgiving car I've ever driven in the mountains, almost miraculously safe.

And so we arrived in good health in Seattle, visited with the Lyons for an evening and the next day drove into Canada on a highway that appeared to lead straight into a spectacular backdrop of Rockies, like the road to the Emerald City in Oz. At the Canadian customs station, we filled out some forms, and the woman behind the counter looked at them and said, "Nineteen *fifty-six?* Wait here."

She summoned an agent who went outside, looked the Citroën over carefully and said, "If this thing breaks down, do you boys know where you can get parts? [Yes.] Do you know how to repair it yourselves? [Yes.] Do you have the means to have it removed from Canada back to the States if it stops running? [Yes.] Do you have any plans to sell it in Canada? [No.]"

He grinned and said, "Well, it looks like fun. These are great old cars. Have a good trip." He waved goodbye as we drove off toward the mountains.

"Well, that was easy," I said. "Last time I crossed into Canada, on my motorcycle in 1967, they kept me up half the night, tore my luggage apart and checked my identity with the FBI. As I recall, it was my haircut they didn't like."

Off we went into Canada, following Highway 1 up the lovely Fraser River Valley as far as Highway 3, an excellent road of sweeping vistas—mountains, pine trees, river rapids and spectacular valleys—that convolutes its way across the southern edge of British Columbia. It was, simply put, some of the prettiest country I've ever seen.

One of the nicest things about British Columbia was that no

Left: The 2CV's seats lift out for handy campfire seating, while table (above) folds out for roadside meals or just feeding the cats. Below: Snagged on a Similkameen River sandbar.

man had yet ripped asunder what nature had joined together. There was civilization in those hills, to be sure, but the towns along the highway were spare, clean, charming and well fitted to the countryside. Miraculously absent were the cheapjack spa-world construction and franchise strips that now form a depressing no man's land around nearly every small Western town on the U.S. side of the border. Somehow the Canadians had kept their towns beautiful, while Americans, only a few miles away across an imaginary line, had largely given up and let everything go to hell. How could this be?

Good judgment? A refusal to go for the quick buck? English tradition? Solid values? Whatever it was, I wondered if maybe Americans shouldn't go to British Columbia and take notes, to see what their towns used to look like, and possibly could again, with the help of bulldozers and dynamite.

Speaking of which, we headed down the lovely Similkameen River Valley and pulled off the road near Cawston. There was a sand bar just below a small bridge, and Chris thought it would be a good place to park the 2CV for photographs. It was. It was also a good place to get stuck in the sand.

We tried everything short of high explosives to get the car unstuck (hunks of old carpet under the tires, tree logs, brush, jacks, prybars, prayer, bouncing, whining, rocking, keening, pushing, powerful oaths, wheezing asthmatically, etc) to no effect. We finally walked to a nearby ranch and found a friendly, good-natured rancher named Larry Winser, who fired up his tractor, hooked a nautical-class chain to the 2CV and extracted it with the same casual effort bears use when batting river salmon out of the water.

"Sorry I yanked it a little," Larry said. "I didn't even feel the chain tighten up."

We stayed at a motel in Grand Forks that night, drank some Kokanee and Kootenay beer and headed down into Idaho the next day, picking up Highway 2 along the south rim of Glacier Park. At Glacier we finally had a chance to camp and to use the 2CV's built-in camper convenience package. We tilted our car seats forward, lifted them out and set them around the campfire, then folded out the hinged table on the side of the truck, using the abundant work space to prepare a delicious meal of Soft Batch oatmeal raisin cookies and Lucky Lager. Talk about really living.

When we got up in the morning, I looked at the map and said to Chris, "We've been on the road a week and we're only two inches east of Medford, latitudinously speaking, and we've got nine inches of map between us and Wisconsin. It's all this

looping around in the mountains, going nowhere."

So it was foot to the floor across Montana and North Dakota. I don't know how long it would normally take to go from Glacier to Wisconsin in a 2CV, but we did it in less than three days, thanks to a howling, relentless wind directly out of the Northwest. With this gale at our backs, the Citroën cruised at an easy 70 mph and averaged 41 mpg. As we approached the Minnesota border, Chris suddenly pulled over and turned the car around on the highway.

"Forget something?" I asked.

"Let's see how fast it goes *into* the wind," he said.

It went 35 mph. Flat out.

So we came about and sailed east before the wind, living on hardtack and limes and whistling hornpipes most of the night, stopping only at a big motel in Mankato, Minnesota for a few hours of sleep before pressing on. Crossing the Mississippi into Wisconsin at Onalaska, we stopped only long enough for a cup of coffee to keep the synapses synapsing and trundled down Highway 14 to Madison. We pulled into Chris's shop late on a Friday afternoon, hit the brakes and bobbed to a stop.

"Well, that was refreshing," I said, climbing out and stretching my legs. "Nice car you got there."

And I meant it. I envied Chris this addition to his small-car collection, and I could see why Russell and Trinka had been content these many years owning no other brand of car.

The 2CV was a great bundle of contradictions that covered all the bases: It was lightweight to the edge of ridiculousness, yet strong and overbuilt where it had to be; it had more dip and sway than any car I'd ever driven, yet cornered like a leech; it was a small, cheap car, but had plenty of leg room and the best highway ride and most comfortable seats this side of a Fleetwood limo; it was low on power, but so much fun to drive you didn't care, outrageous but socially responsible in its consumption of resources, a stridently ugly collection of shedmetal at first glance that became charming and maybe even beautiful with familiarity; finally, it was an immensely practical car that exuded a sort of edge-of-reality beatnik nihilism. In France the 2CV was a badge of working-class pride, while on this side of the Atlantic it had powerful cachet among the tuned-in educated semi-idle Bohemian rich. It was a rough-cut jewel whose facets reflected whatever the owner wanted to see.

The Citroën 2CV was perhaps the first, last and—so far—only compleat car ever made. It was also a car you might have built yourself, but hadn't.

Honest.

The Wee

Each year, at the end of summer, vintage-car race drivers, collectors and enthusiasts from around the world set their focus on the Monterey Peninsula in California and the three-day carnival of automobilia that has come to be known as...

The Races

Is it racing or is it Memorex?

BY PETER EGAN

MONTEREY IS A place where you wake up in an English frame of mind. The days start out dark and gray, with enough fog and low coastal clouds to make Aston owners think they've died and gone to Feltham. But if the mornings have a North Atlantic feel, the afternoons are Mediterranean. Clouds burn off, the sun comes out and the landscape goes Italian, with hills the color of straw hats. Hard sunlight glints off the Borrani wheels and red paint of Ferraris and Alfas, which suddenly seem to be in their element. Somewhere in the transition between cloud and sun, the German, French and American cars presumably feel right at home. There's weather for everybody.

Cars, too.

Good Lord, the cars.

This year it was Aston Martin's turn to be the featured marque, and they did it up right. Fans who arrived on Friday morning walked through the paddock gate to discover that Aston had

kend

PHOTOS BY JOHN LAMM, RICHARD M. BARON & SCOTT DAHLQUIST
SKETCHES BY LEO BESTGEN

■ Standing in front of Aston Martin's re-creation of the Le Mans pits, you could almost hear the *whoop, whoop* of the engines and the roar of the crowd. The illusion was complete right down to the mannequins dressed as gendarmes and overall-clad mechanics.

built a full-size replica of the Le Mans pits, complete with dummy mechanics dressed in white BP coveralls, and pit signs bearing abbreviations of names like Shelby, Salvadori, Trintignant, Frère, Moss and Fairman.

More remarkably, those very drivers were milling around, in person, amongst an amazing collection of DB and DBR Aston Martin racing cars, all in silver-green Aston livery and angled as if for a Le Mans start. Sir David Brown (the original DB) was also there, as were Aston President Victor Gauntlett, Aston's part-owner Peter Livanos, Phil Hill, Innes Ireland, Tony Brooks and Bib Stillwell. Even Ted Cutting, the designer responsible for the whole line of DBs and DBRs, had made it to the reunion.

It looked as if some kind of selective hypergravity had drawn in every living person and thing that was ever important to Aston Martin and plunked it down at Laguna Seca. Only the much-missed General Manager John Wyer, who died at his home in Arizona this year, and the late Reg Parnell, Aston's team manager from the Fifties, were absent.

Among the cars in the Aston pit-row lineup was the DBR1 that Carroll Shelby and Roy Salvadori drove to their 1959 victory at Le Mans, in one of the sweetest wins in Le Mans history, coming after a decade of frustration and near-

■ Above, a beautiful 1957 Aston Martin DBR2 at speed, driven by Mike Salmon. At right, two cars representing the full range of engine displacements: a 1912 Bugatti-designed Bébé Peugeot with 750 cc, and a 1911 Fiat S74 GP with a staggering 14 liters!

misses by the Aston Martin team. Having the car and both drivers together, 30 years later, was the sort of celestial convergence only historic racing seems to manage.

Other traps were scattered around the paddock like so much flypaper. I turned away from the Astons to see a pair of 1958 Vanwall GP cars from the Miles Collier museum in Florida, nearly backed up into Dick James's 1959 ex-Brabham Cooper F1 car, careened off a bank of Bugattis and found myself stumbling toward a green Morgan 3-wheeler belonging to Spence Young. Spence fired up the Matchless V-Twin and gave me my first-ever Morgan trike ride. We cruised through the paddock, visiting people and watching the external rockers twitch on either side of the radiator shell.

When we got back to Spence's parking spot, Jay Leno (*Tonight Show* guest host, Morgan trike owner and genuine car/bike nut) walked up to look at the 3-wheeler. Leno said he was driving his own Morgan 3-wheeler recently when the spare tire fell off the back, without his noticing. A motorist drove up alongside and shouted, "Hey, you lost a wheel!" Leno just smiled and waved and continued driving. "Jeez, that old joke again," he mumbled wearily to himself.

Down on pit row, my old friend Larry Crane was getting ready to drive Steve Earle's Jaguar C-Type in practice. He told me they'd put on new rear tires with too much tread and the rear end was sliding all over the place, hurting their chances for a win in Group 4. "But it's just vintage racing," I said, "so it doesn't matter where you finish, right?"

Larry looked at me and smiled wryly. "Yeah, right," he said.

Dick James was likewise worried about his chances in the upcoming exhibition race for later Formula 1 cars. Dick is a photographer who owns and races the ex-Brabham Cooper I mentioned, plus an ex-Denny Hulme 1970 McLaren M14. When I walked up, he was staring forlornly at the McLaren.

"When I built this thing, I wanted reliability and I thought 480 horsepower would be enough. Well . . . it's not."

There was the old conflict again, finding a balance between history and the need to compete. Steve Earle, who organizes the Monterey Historic Automobile Races, has tried over the years to de-emphasize wheel-to-wheel competition as much as possible. Drivers can be disqualified and barred for metal-to-metal shunts and overly aggressive driving, or black-flagged for various off-track experiments. Until recently, trophies were given for last place and winners were largely ignored. (Earle stopped this practice because back markers were getting on the brakes and engaging in weird tactics to finish last on purpose.) Now the silver cups are handed out for a finishing position chosen at random—this year all 9th-place finishers got a trophy.

So the spirit of the Monterey Historic has always been that of the Living Museum, a chance

to bring out wonderful old sports and racing cars and drive them at high speed, while still preserving them for posterity. Beneath that fine ideal, however, lurks the human need to compete. Most of this manifests itself in flawless car preparation, which has rescued hundreds of old, neglected race cars from oblivion and returned them to their former glory. The other part shows up as good hard driving.

Somehow, the Monterey Historic organization (General Racing Ltd.), with a little help from the drivers, has been able to strike an almost perfect balance between these forces. The result is a kind of automotive heaven on earth, where the crowd gets to see some of the most beautiful cars ever made, driven very fast

■ **From left to right, legendary men from Aston Martin's past: Carroll Shelby, Roy Salvadori, David Brown, Stirling Moss and Jack Fairman.**

indeed rather than stored in a museum. It makes for a good show.

Where else, for instance, could you have seen Stirling Moss with his old forceful, decisive flair, driving a DB4GT and somehow elbowing his way to the front of a pack of 25 Astons, Ferraris and Corvettes in Turn 2 without actually touching any sheet metal? Unfortunately, this race, second to last on Saturday, did produce some bent sheet metal—and shredded fiberglass—when one 1957 Corvette ran into the back of another, touching off a four-car tangle that also damaged a short-wheelbase Ferrari 250 and an Aston DB4GT. There was a lengthy pace-car-led parade as the expensive debris was cleared by wreckers.

Above, Murray Smith in an Aston Martin DB4GT has Stirling Moss and a host of other Aston drivers breathing down his neck while negotiating Laguna Seca's Corkscrew. Below, jewel-like was the way to describe the engine-turned aluminum body on this 1924 Delage.

Then there was the sight of a whomping 1953 Kurtis 500KK, owned by Robert Lamplough and driven by Michael Sheehan (whose restoration shop is about two blocks from our office), smoking a field of DB3Ss, Allards, C-Type Jags and various specials with good driving and big gobs of loud horsepower. Our friend Larry Crane finished a smooth, fast, spin-free 5th in this race, despite his slidy rear tires.

Sunday gave us the sight of Dan Gurney and Innes Ireland dicing with commendable restraint in those two matching green Vanwalls, arguably the most beautiful front-engine F1 cars ever built—arguable because Peter Giddings' stunning Maserati 250F was also out there, wailing and drifting, shark-like.

You could go on forever, remembering the GT40s, Sprites, Cobras, Alfas, Ford T specials, MGs, Lancias and Lotuses. There are no bad races at the Monterey Historic, no dull races, no moments of boredom. Between races there are cars to look at, and when you tire of just looking, there are engines to hear idling and revving, and when the engines shut down, there are people to talk to—people Steve Earle described in this year's race program as "true enthusiasts, lovers of mechanical marvels, hero worshippers and curators of automotive history."

Well said.

Continued from page 79

> ## "*I am a former Grand Prix pilot,*" *Innes explained patiently,* "*and I don't throw my beer glass onto the bloody track. You can trust me.*"

Innes later apologized to us, but I thought he was right. Maybe if more of us made a fuss, we'd get real cream for our coffee instead of whitener, and we wouldn't be scraping jelly out of little plastic containers onto our toast. You do get tired of the glass-flingers of this world winning every round.

In practice on Friday, Senna went even quicker. As other drivers struggled to get into the 1:28s or 1:29s, Senna just kept going faster and faster until he had posted the incredible time of 1:23.99, nearly 1.5 sec ahead of teammate Prost, who was himself more than a second faster than Berger's Ferrari. It looked as though Monaco might be set to follow the season pattern by yielding again to one or more of the Four Irresistible Forces: Honda's engine, McLaren's chassis, Senna's sublime speed and Prost's masterful combination of skill and good judgment.

And so it turned out.

I awoke on Sunday with the kind of adrenaline-inspired alertness that overwhelms even too much Côtes-du-Rhône, drunk late the night before. Grand Prix cars emit an electrical charge on race day that makes sleep impossible. I hiked down to the hairpin with my camera gear, while Barb went up on the swimming pool terrace of Loews, which provides an excellent view of both the hairpin and Portier, along the water.

Now I don't know about you, but I happen to think the best moment on any race day is when the cars first take to the track. Particularly if they are Formula 1 cars. You wait at a corner listening to engines warming up in the distance, and then they get louder and louder until the first car suddenly materializes in the corner like an explosion of color and noise, and dozens more come snarling in behind it. You really do feel as if the tigers have been let out of their cages and there is something deadly loose on the circuit. There is an element of pent-up rage and hostility in the way F1 cars hunt and weave when they are warming up, as if they're angry from being held too long in the pits. Suddenly Armco doesn't seem like such a bad idea.

The race started and Senna led. And led and led. While Prost struggled to pass a determined, hard-driving Berger, and Mansell drove his heart out hanging onto both, Senna was off in some other world, circulating like the very embodiment of speed as imagined in an opium dream. He came whistling through the turns, never an inch off the line, those fat tires whipping back and forth and just grazing the guardrails, as if daring them to be even slightly in the wrong place. It was an unreal, ethereal display of driving talent, of wheels put just right and of margins split over and over again into infinitely smaller parcels of space.

"That kid can really drive," I said to no one in particular.

Senna's lead stretched from 5 seconds to 10 seconds to 30, and it kept growing. I walked up to the long curve leading into Casino square and watched the cars drift through, flying over the hill and converting lightness to traction with steering wheels and brakes. While most drivers tiptoed through the curve under some sort of steady-state throttle, Senna did it with half a dozen short bursts of acceleration that kept the car alternately drifting and scrubbing off speed, or so it looked to me.

Senna's lead went to an unbelievable 52 seconds, and I began to wonder why he wasn't slowing down. Did he think that he might actually lap Prost by the end of the race? Was he in such a groove of euphoric speed that he couldn't back out of it without disturbing his perfect rhythm? Or was he just driving as fast as he could go because that's what he *does*, like Nietzsche's eagle, helpless to be anything but an eagle? Whatever it was, he kept up the relentless pace, finally backing off only a little, under team orders.

And then, suddenly, he was gone. No flying yellow helmet. No Senna.

Prost, having finally got by Berger, was in the lead, and I learned that Senna had hit the rail at Portier. Nothing nasty, just a good, disabling clout of the wings and wheels. He was out of the race. As was Mansell, who was punted out of a brilliant 3rd place by Alboreto.

I walked down the hill toward the start/finish line and got there just in time to see the race end, the crowds storm over the barricades, and to watch Prost, Berger and Alboreto climb the victory platform, shake hands with Prince Rainier and spray champagne on the fans.

It was a well-earned victory by Prost, a great fight for 2nd by Berger, and I'll have to watch the tapes a few more times to see if Alboreto played fair with Mansell. Prost now has 24 World Championship points, Berger 14 and Senna is ranked third, with 9.

There was a lot of post-race discussion of Senna and his apparent reluctance to slow way down and cruise home for an easy victory. In Senna's defense (if he needs any from the likes of me), I would say it probably *was* his slowing down, and the attendant loss of concentration, that caused the crash. In any event, his display of driving prowess certainly made life more interesting for the paying crowds. Winning is supposed to be the end-all in a modern Grand Prix, but there's a lot to be said for pure, uninhibited speed. As we amateur sports car racers used to say in mock European accents just before climbing into our cars, "It is good to be young and go very, very fast."

Indeed it is.

On Monday morning, Barb and I checked out of the hotel and I descended into the bowels of the earth to retrieve our car from the underground parking garage. The parking fee was $107 for the week, but at least they let us stop and get out of the car this time, so Monaco has come a long way since 1973. So have we, of course, graduating from a Simca 1000 to a Fiat Panda.

To those, like my friend George, who have it in mind to go to Monaco, I would say by all means do it. Preferably before you die. And bring money.

Winner Prost (center), Berger (left), Alboreto, Moët and company.

NORTH AMERICAN MUSTANG P-51D

It's the fastest Mustang we've ever tested, but is it a match for the dreaded Messerschmitt?

BY PETER EGAN

PHOTOS BY RICHARD M. BARON

ROAD & TRACK HIGH TEST

Lt. RICK BARON and I had just returned from the USO club and a game of darts with some of the Limey flyboys when Reynolds showed up in the wing commander's "borrowed" Jeep with a bottle of scotch in one pocket of his raincoat and a tinned fruitcake in the other.

"No boiled brussels sprouts tonight, boys," he said. "We can trade this hooch for some real eats." Reynolds is seeing an English nurse in the WAAFs whose old man is some kind of bigwig in procurement, with a title that goes back to King Arthur and a house that makes the Pickford and Fairbanks place look like it's strictly from hunger. He can get us everything but home.

We went into the ready room, where the Old Man, Col. Bryant, was dressing down one of the new guys for having used a dangling participle on the radio over Düsseldorf. Seems he also messed up "who" and "whom." Big trouble. It's not easy flying in a motorjournalist squadron, and the new kid may not make it. I heard him say "orientate" out on the strafing range.

We were about to slink off toward the officer's club bar when the Old Man called us into his office for a briefing. "Right now," he growled, in his usual genial manner.

"Here's the story, gentlemen," he said, leaning on his desk and squinting at us through a cloud of cigar smoke. "The top brass has sent down word there's a new Mustang on the way, a P-51D. It's got everything; speed, range, drop tanks, six Brownings. You three have been assigned to go stateside and look at the new bird. Some of the other units have had it since 1944. I want to know if we need it."

We all looked at one another. Art Officer Baron raised an eyebrow. "Seems to me, Sir, we're doing okay with the Mustangs we've got now," he said. "We've got the GTs with the HO 302 and the handling package. What more do we need?"

Baron has 21 kills, having just smoked a 325i last Monday in a roundabout at the Biggin Hill exit. He likes the GT.

"There's nothing wrong with the basic Mustang GT," Col. Bryant said, "but it's a car, for God's sake, and the P-51D is an airplane. Our B-17s are getting shot out of the sky because we can't give them any air cover. Our current Mustangs can drive as far as the cliffs of Dover and then they have to turn around. We want a Mustang that can go to Berlin and back. So pack your gear, gentlemen. You're leaving tonight for Mojave, California."

As we stood up, he said, "One more thing. The intelligence boys have reconstructed a crashed Messerschmitt and we're sending it along for comparison. Fly the P-51, look at the Messerschmitt and tell me what you think. I want to know if we have a chance against this thing."

We flew stateside in a homebound Liberator, did the usual buzz under the Golden Gate and found ourselves in Southern California. The guys dropped me off at Chino Airport, where the P-51 is kept, and then headed for Mojave Airport, up in the high desert, with our test equipment in the back of a Corvette ZR-1. Intelligence thought this was the only car

that had a chance of keeping up with the P-51 on its takeoff roll, to measure acceleration. (As usual, Intelligence didn't have quite enough.)

I was to meet the Mustang owner, Elmer Ward, at his hangar in Chino and fly up to Mojave with him. There we would rendezvous and conduct our secret comparison test with the Messerschmitt.

Elmer Ward drove up and we shook hands. Ward is a slim, energetic industrialist in his 60s, a Cal-Tech-trained engineer and owner of his own manufacturing firm. He also owns a company that builds new Mustangs from a vast supply of surplus parts he keeps at Chino. He

■ Instrumentation is extensive but cowl height is a bit imposing—all the better for slouching to avoid eye-to-eye contact with the enemy. What looks like a window crank at the right of the cockpit locks the canopy down; a lever just ahead blows it free in an emergency. Wheel to the left of the seat is the elevator trim; large knobs above it are rudder and aileron trims.

"Feats of derring-do like dragging my wingtip on the desert floor at 300 miles per hour will strike terror into the hearts of Luftwaffe pilots," thought our man, grinning impishly.

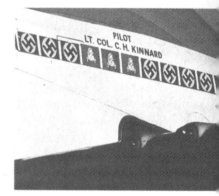

rolled open the hangar doors and showed me his own Mustang, a P-51D built in 1945, a real beauty.

As we looked the plane over, Ward gave me a little history. This particular P-51 was built in Inglewood, California, flown to England during the last months of World War II, and lent much later to Universal Studios for the 1956 movie *Battle Hymn*, with Rock Hudson. A man named Ascher Ward (no relation) bought the Mustang off the back lot in 1970 for $3500 and restored it to flyable condition. Elmer Ward bought it in 1975 and did a complete restoration in 1980.

The airplane is now painted in the colors of Col. Claiborne H. Kinnard, the famous ace (27 kills) and commanding officer of the 4th Fighter Group of the 334th squadron, 8th

Air Force in England. Kinnard survived the war and died at his home in Franklin, Tennessee in 1967. Ward painted the P-51 in Kinnard's colors out of admiration for Kinnard himself and for the 4th Fighter Group, which began life as the three original Eagle Squadrons, formed of American volunteers who fought for England before the U.S. entered the war. Ward actually flew the Mustang to Cleveland in 1976 to have Kinnard's *Man O' War* script painted on the nose by Don Allen, the original artist.

Enough history. Time to go flying and take some road test notes for the Old Man.

Ease of entry? Fair. You step on the tire and landing gear strut, climb up on the front of the wing, crank back the canopy and just step in, if you are the pilot. As a passenger, I

had to slither back under the canopy into a rear seat installed where the 85-gallon fuselage gas tank used to be. The plane still carries 180 gal. (usable) in the wing tanks, enough for about 800–900 miles of cruise. Under the wings and fuselage are attachment points for the auxiliary drop-tanks that extended the P-51's range deep into Germany, making life good for Allied bomber crews and bad for the Luftwaffe.

Parachute harnesses first, then seatbelt and shoulder straps. Elmer says if you bail out of this baby you pull a canopy release lever, duck low so the departing canopy doesn't take your head off, crouch on the edge of the cockpit and leap hard toward the right wingtip. This way, with luck, you will clear the tail so it doesn't break both your legs as it passes by.

A booster cable from an auxiliary power unit is hooked to the belly, Elmer calls "Clear!" and hits the starter button.

When the prop begins to turn, there's something startling about seeing those four huge blades jerk so quickly away from rest, the sheer mass and diameter of them—11 ft. 2 in. of aluminum alloy.

Two giant puffs of smoke explode from the six straight stacks on either side of the nose, and the supercharged 1649-cu.-in. V-12 (the displacement of 4.7 Corvette engines) is running and muttering like a heavy smoker getting up in the morning and clearing his lungs. Even at idle, there's an undertone of menace in the deep exhaust note. It says, "This is a warbird, pal, and engines don't sound this way unless they are going to war." You feel that the Beast has been awakened, and he's not happy about it.

We taxi between hangars, watching our wingtips and looking down on the roofs of cars. Five minutes' wait in the runup area while 21.2 gal. of 60-weight oil and 16.7 gal. of coolant warm up, and we taxi into position and hold. Throttle comes up, brakes are released, manifold pressure climbs to 50 in. (they used 61 in the war) and the Mustang thunders forward, trying to torque its tail sideways against Ward's steady rudder pressure. The tail comes up and we are flying.

Climb is so easy and lazy, it's almost an anticlimax; 200 mph, 2000 ft. per minute with the nose barely over the horizon. The great sink that contains Los Angeles drops from beneath us, and we are headed for the snow-dusted San Gabriel Mountains. A 240-mph (indicated) cruise at 2150 rpm takes us over the peaks at 9500 ft. The engine has the mellow, rhythmic shuffle of a sewing machine made from locomotive parts. The Mustang is over the high desert now, trimmed nose low for a wonderful view of ground and sky. Forward vision is excellent. The airplane has assumed its natural gait, which is somewhere between prowling and ranging. It cruises along head down, purring, but full of latent ferocity, no doubt wish-

ing it had a train to strafe.

I lift the headphones experimentally away from my ears for a second and am assaulted by the loudest mechanical cacophony I've ever experienced. It's like . . . like what? Like having a stethoscope on a tin roof during a ball-bearing storm. The headphones go back on and I try to refocus my eyes. I won't do that again soon.

"Some aerobatics?" Elmer asks over the intercom.

"Sure!"

The brown/blue horizon tumbles and tilts. Eight-point rolls, 4-point rolls, barrel rolls, right and left knife-edge, Cuban 8 and a wonderful bank reversing maneuver called a John Derry Turn that I've never done before. Elmer Ward does them all with a velvet touch and confident precision that would reduce even my old aerobatics instructor to an appreciative silence. Twisting and turning over the desert floor, the Mustang has the lazy, muscular confidence of a Great White shark, casually deadly and in no particular hurry. It feels stable, powerful and solid.

We bank off toward Mojave airport, not far from where the space shuttle lands at Edwards, spiral down and swing in for a fast but gentle 3-point with all 12 pipes crackling like a string of Black Cats on the 4th of July. We taxi, canopy cranked back, breathing clouds of heady rich 100-octane avgas fumes. Elmer parks and shuts her down. The sudden silence is deafening.

"This is a wonderful airplane," I say, stating the obvious.

Ward agrees. "I think it's the best fighter of WWII. It's strong and simple—just four big longerons with

skin riveted to them and a big engine in the front. It was also the cheapest fighter of WWII, $50,000 apiece, without engine."

We climb down from the wing and Reynolds is there, with his Corvette and test gear. The Messerschmitt is arriving tonight, Reynolds tells us. We test in the morning.

At sunrise, our friend Paul Prince arrives with his two sons, Austin and Zeke. They have brought their 1953 Messerschmitt KR-175 over from Santa Barbara in an unmarked white rental van. Smart. We unload the little ship, and, frankly, she doesn't look like much next to the Mustang, but you never know. The Messerschmitt is powered by a 9-bhp Fichtel & Sachs 2-stroke single, the Mustang by a 1400-bhp Rolls-Royce Merlin V-12. Is this the Messerschmitt that Col. Bryant had in mind? It doesn't even have wings.

Testing is somewhat inconclusive. While the Messer is more agile through the cones, the P-51 simply blows them away with prop blast. Which technique is better? I'll take the P-51.

Lateral acceleration? We didn't have access to a skidpad at Mojave, but the P-51 generates the usual 4 to 5 g's in simple aerobatic maneuvers and is structurally rated far beyond the ability of humans to withstand g-forces and remain conscious. I doubt the 4.00 x 8-in. tires on the KR-175 can equal this feat. The Messerschmitt produces no blackout, even in hard turns. The KR-175 is also slow, with a top speed of about 48 mph, versus the Mustang's 428 mph.

About all the Messerschmitt really has going for it is good fuel mileage (102 mpg vs. 5.0 mpg for the P-51),

■ Fear of being sucked into the P-51's prop motivated our man to post what is possibly the fastest slalom time ever achieved by a car powered by a Fichtel & Sachs 2-stroke.

NORTH AMERICAN MUSTANG P-51D

Takeoff roll	**1500 ft**
0–40,000 ft	**25 min**
Top speed	**428 mph**
Dive speed	**505 mph**
Vertical accel	**8.0g**
Landing roll	**1800 ft**

PRICE

List price,
FOB Inglewood, Calif ... **(1945) $50,985**   Price as tested ....**est $75,985**
Price as tested includes std equip. (bulletproof windscreen, ejectable canopy, six Browning 50-caliber machine guns, drop tanks, cruise control), Packard-Merlin V-1650-7 engine (est $25,000).

ENGINE

Type	two-stage supercharging, inter & after coolers, alloy block & head, **V-12**
Valvetrain	sohc 4-valve/cyl
Displacement	1649 cu in./ 27,022 cc
Bore x stroke	5.40 x 6.00 in./ 137.2 x 152.4 mm
Compression ratio	6.0:1
Horsepower (SAE):	**1400 bhp @ 3000 rpm**
	1720 bhp @ 3000 rpm
	(during war emergency; 5 min. max)
Bhp/liter	51.8
Maximum engine speed	3240 rpm
Fuel injection	Bendix PD-18C1
Fuel	premium leaded, 100 pump oct

DRIVETRAIN

Transmission	**constant speed propeller**
Type	Hamilton Standard 4 blade
Diameter	11.2 ft
Pitch	variable, 23–65 deg
Actuation	automatic, hydraulic
Cruising speed @ rpm:	310 mph @ 2300 rpm
Maximum speed	
(level flight)	428 mph @ 17,000 ft
Maximum speed (dive)	505 mph

CHASSIS & BODY

Layout	**front engine/front drive**
Fuselage & wings	aluminum skin on four longerons, main & rear wing spars, 25,000 rivets
Brakes, f/r	**7.5-in. discs/none,** flap assist
Wheels	cast alloy; **14 in. f, 5.5 in. r**
Tires	Uniroyal Aircraft, **27 x 10 in. f;** McCreary, **12.5 x 4 in. r**
Steering type	**(ground) tailwheel, (air) 10.4-sq-ft rudder**
Suspension, f/r:	retractable; **pneumatic struts,** oil damping

HANDLING & BRAKING

Vertical accel	8.0g positive
Balance	affected for weeks
Speed thru 700-ft slalom	unknown, prop blows all the cones over
Minimum stopping distance	
From 87 mph (landing speed)	1800 ft
Overall brake rating	nothing broke

GENERAL DATA

Curb weight	**8320 lb**
Test weight	**8520 lb**
Weight dist, f/r, %	**92/8**
Wheelbase	16.3 ft
Track, f/r	11.8 ft/0 ft
Length	32.2 ft
Width	37.0 ft
Height	8.7 ft
Trunk space	7.9 cu ft per wing or six machine guns & 2080 rounds

INSTRUMENTATION

700-mph speedometer, 4500-rpm tach, oil press., oil temp, fuel press., coolant temp, induction air temp, +8/−4g vertical accelerometer, manifold press., dual RMI, suction press. for artificial horizon, artificial horizon, 40,000-ft altimeter, turn & slip indicator, rate of climb indicator, horizontal situation indicator, oxygen system press., compass, dual fuel level indicators

FUEL ECONOMY

Normal flying (300 mph)	5.0 mpg
EPA rating	4.4 city recon/
	4.4 highway strafing
Cruise range	900 miles
Fuel capacity	184 gal.
Oil/filter change	125 hrs
Tuneup	100 hrs
Basic warranty: unlimited missions or VE Day	

ACCELERATION

Runway time to speed	
0–30 mph	Corvette ZR-1-like
0–60 mph	the Vette is doorless
0–80 mph	no contest
0–100 mph	almost airborne
Rotation	1500 ft @ 110 mph

Time to altitude	Minutes
0–10,000 ft	3.7
0–20,000 ft	7.0
0–30,000 ft	12.6
0–40,000 ft	25.0
Service ceiling	41,900 ft

INTERIOR NOISE

Idle	95 dBA
Maximum, takeoff	125 dBA
Constant 240 mph	115 dBA

Subjective ratings consist of black and white with no shades of gray. This is World War II, pal!

---

easy entry and parking, and low maintenance costs (a Merlin V-12 rebuild costs $45,000; the Fichtel & Sachs 175 could probably be rebuilt for around $100).

Measuring 0–60 acceleration on both turned out to be a bust. The Messerschmitt won't go 60, and our Corvette couldn't keep up with the Mustang. When we took off to leave Mojave, the Corvette ZR-1 and its 5th wheel lined up at the P-51's right wingtip, hoping to pace the Mustang and measure its performance. The smart money said the Corvette would be loafing until 60 or 70 mph. The smart money was wrong.

Elmer Ward ran the Merlin up to 3000 rpm and 30 inches of manifold pressure, released the brakes and shot down the runway, building to 50 in. of pressure. The Corvette never had a chance. It stayed with us until about 40 mph and then disappeared into the Mustang's twin mirrors like something tied to a post. We swooped into a climbing 180-degree turn at 200 mph and seconds later were flying knife-edge, looking down our wingtip at the Corvette's roof. The Messerschmitt stood by the side of the runway, growing smaller and smaller in the vastness of the desert until it was lost among the rocks and tumbleweeds.

It's easy to see now why the P-51 has been such an overwhelming success in the other squadrons. I think the Old Man is going to like this new Mustang. I think he's going to like it a lot.

The Messerschmitt boys are going to wish they'd voted for Roosevelt. ⬣

Test Notes . . .

■ Absolutely the most powerful Mustang we've ever tested, the P-51D's prodigious 1400 bhp produces nary a chirp from its tiny tires off the line. Would likely better 110 mph if not for wings mounted upside down causing it to fly.

■ Absolutely the loudest Mustang we've ever tested, the P-51D's going to collect plenty of fix-it tickets in hospital zones. Speaking of micro-surgery, the Mustang's radiator fan can make for a nasty nick. Ouch!

COMING TO INDY

When Luyendyk, at last, in the Brickyard bloomed

BY PETER EGAN

WHAT! NEVER BEEN to Indy before? Then you should have stayed away," the man says. He's an older gent, wearing clip-on sunglasses and a Speedway cap with a winged-wheel emblem on the front. "You don't ever want to come to Indy for the first time. I did it myself in 1951, and I've been back every year since. It gets in your blood, this place."

Never been to Indy before. The words sit there like a flat disclosure of missing spark and gumption. Almost like telling people you never learned to drive a stickshift, tried a cigarette or tasted whiskey. It's hard to admit, especially at the age of 42, but this May was my first trip to the 500.

And me a Midwestern kid too, growing up with scrapbooks full of USAC drivers with reckless grins and polo helmets, haunting the local circle tracks in summer, building models of Calhoun and the Boyle Special and staring at them for hours.

Always celebrating Memorial Day weekend with those strange, uniquely American ceremonies that blended more images of death and rebirth than the complete works of Walt Whitman: high school bands marching to small-town cemeteries, volleys of rifle fire over new-blooming lilacs, veterans' paper poppies worn in lapels on sadly beautiful spring mornings, *In Flanders Fields* read in a quavering schoolgirl's voice while small American flags fluttered next to granite headstones. Taps played and echoed mournfully from somewhere down in the woods.

Later a drive home with my dad, and I'd sit in the Buick in front of the house listening to the race coverage on the radio. Foyt, Sachs, Jones, Rath-

mann, Bettenhausen; tough guys and heroes, somehow linked in courage and resolve with the soldiers at Normandy or Shiloh or the Ardennes.

It all fit together, Memorial Day and the Indy 500, a celebration of spring with pungent overtones of danger. But instead of rifles and cartridge smoke, it was Offys and alcohol fumes and the brightly colored roadsters with their snap-in upholstery, big steering wheels and bobtail gas tanks. Bowes Seal Fast Special, Leader Card 500, Dean Van Lines Special, narrow Firestone tires and a whole grid filled with the talented survivors of a hard craft, still with us because they had *not* gone over the fence at Trenton or Terre Haute, or if they had, they'd landed right, maybe more than once, and had that blue light of invincibility and immortality about them.

And when the blue light failed, it was tragic, yet undeniably enlarging to the other drivers, who were now, more than ever, survivors and heroes. It was trial by combat, set in a ceremony of martial music, baton-twirlers in golden swimming suits, prayers and patriotic songs, the air heavy with summer's first haze. Let the Teutons and the Druids have their hilltop rituals of the new season with oak trees and sunrises. Next to Indy, their medicine was weak.

Foyt takes the checker! Hot damn, you say. Summer's here. The teacher let the fools out. We're alive again.

This and more was all Indy and Memorial Day to me, but for some reason I'd never made it to the race. Maybe it was just arenaphobia—the

■ **Beating the odds and besting his rivals, Arie Luyendyk and his Lola-Chevy proved a winning combination.**

fear of too many fans competing for too few seats and parking places. Or maybe a fear of finding something too powerful or not powerful enough after all these years.

Whatever the reason, there comes a year when it's time to go. Part of Nature's clock perhaps, like the unheard signal for lemming migration. You call up the travel agent, book a flight, and suddenly you're on a commuter twin Beech 1900C out of Wisconsin, flying in for practice and qualifying.

Two hundred twenty-five miles per hour, the pilot says, in bright sunlight, just over a near-solid layer of Midwestern stratocumulus gloom. You watch the flat lake of cloud stream by and think this is how fast Fittipaldi and Mears and the boys will be going, lapping Indy. Only with walls. The plane drops through the cloud and we are over Indianapolis, with a shaft of sunlight hitting the city, right out of a William Blake woodcut. The weather looks unpromising for Pole Day.

A rental car gets you a short distance north to the track and you come upon it almost by surprise, scanning the horizon for something huge like the Colossus of Rhodes, and instead you discover, at the corner of 16th and Georgetown Road, a low edifice that looks at first like the backside of a high school football stadium.

Then you see that it goes on and on and that seats and balconies rise out of the surrounding Hoosier flatland and you are on the periphery of something unusual, something there's maybe only one of on earth. A sprawling, huge race track with seating for half a million. The stands flank upward and outward like the

INDY

petals on a great rectangular flower. The seats go on forever.

And to get good seats, a man on the plane told me, you have to wait for someone to die. Order the $18 seats the first year and gradually seniority will get you into the $25 seats, then $50. As a generation passes on, the $100 seats will eventually come your way, sure as liver spots. I wonder if $100 seats are nature's way of telling you to slow down, or maybe schedule a prostate examination.

There's always the infield, of course, where some watch and some drink, and others watch others drink. Drive your car on in, through the tunnel right under the track and park somewhere. Anywhere. The infield's huge.

I walk to the credentials office and mention to a guy in line that this is my first time here and he practically adopts me for the morning. Albert Wong is his name, a polite, low-key fellow who shows me all around and points out the key sights. Turns out he knows everybody. They say, "Hi, Albert!" Even George Bignotti, who cruises by in a golf cart. Seems Al-

bert used to be with Teddy Yip and Theodore racing, and now does computer work for Dan Gurney. Gasoline Alley—which is not just an alley of fuel pumps, as I thought, but a complex of garages for the cars—is teeming with people who nod and smile. It's as relaxed and friendly as the paddock at a small-time sports-car race. Except the nodders and smilers have names like Rutherford and Rahal and Gilmore.

"Are you having a good time?" That's the question everyone asks when it's revealed you're a first-timer here. And they really want to know.

You get the feeling it's a huge private club, Indy, but the other members want you to belong. The true believers are like Transylvanians or the hill tribes of the Andes, watching you sample their goat cheese and their strange medicinal drink, watching to see if you like it as much as they do. You try the strong drink and smile and the village smiles with you. You get a slap on the back and you're in. Or at least you have a foot in the door.

Old Indy fans, like my friend in the Speedway cap, will tell you right off it's addictive, that you'll be coming back to Indy for the rest of your life.

Others aren't so sure. A photographer in the pressroom tells me, "You get tired of this Hoosier thing. The hoopla, the buildup, the endless ceremonies. The officials with their badges. It's just another race, for

God's sake. Same as Phoenix or Milwaukee."

Is it?

You look around and there are thousands of people in the stands, two weeks early, just to watch the cars practice. Is this just another race? Maybe the hoopla is self-fulfilling, you think. In believing something is important, we make it so, and thousands of fans think practice and qualifying are important. Better withhold judgment and see for yourself.

Practice time. Teams tow the cars out backward with straps and Sears garden tractors, while golf carts bring the drivers and team managers down pit row. I step out of the way of a cart carrying Emerson Fittipaldi, Rick Mears, Roger Penske and Teddy Mayer. The crowd applauds at the sight of all that condensed talent. Drivers get into cars, cameras click and engines whoop to life, filling the air with sweet methanol fumes, like a great jug of white lightning suddenly uncorked.

A.J. Foyt walks down pit row and a swell of cheering rises from the crowd. "They love A.J. here," another reporter tells me. I'm pretty happy to see him here myself. He drove in the first race I ever listened to, 32 years ago, and he's still looking healthy at age 55. Some people say he's slowed down in recent years. Maybe. But later in the week he'll qualify Jim Gilmore's Copenhagen Lola/Chevy in the middle of the third row, at 220.425 mph, leaving about 25 younger, theoretically hungrier drivers parked behind him. Someone will ask Foyt where he found the extra burst of speed and he'll say, "My foot."

More applause and Mario Andretti appears, carrying his helmet. Mario has a face right off a Roman coin, dignified and worldly-wise. The field is full of Andrettis—son Michael, nephew John. There are also a

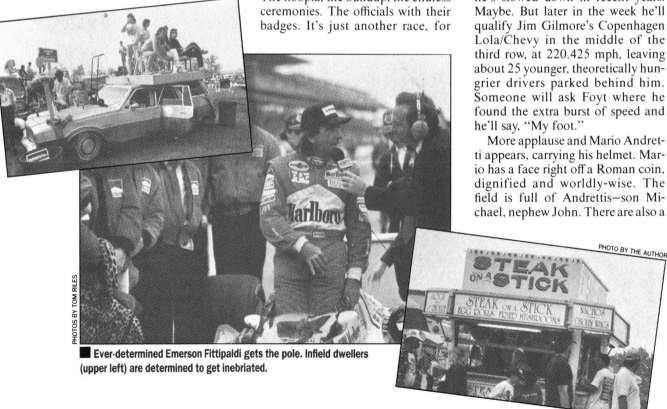

■ Ever-determined Emerson Fittipaldi gets the pole. Infield dwellers (upper left) are determined to get inebriated.

couple of Bettenhausens, a young Brabham, a Vukovich and the father/son Unser duo. With all this nepotism, no wonder I can't get a ride.

The sun shines on practice the day before Pole Day, and the speeds are amazing. I watch from Turn 1 as Fittipaldi burns up the track, faster with each lap until he turns a record 227.181 mph, then Little Al goes out and does 228.502 for a new unofficial practice record. Mears is right behind them. The slower cars are visibly driving around the track, but these front-runners appear almost to be beaming past, like reflected light hitting mirrors on the corner walls.

Then, speaking of walls, Jim Crawford suddenly smokes his tires right into the Turn 1 concrete, bounces and leaves the ground, his Glidden Paints Special rotating slowly in midair like a show car on an invisible display platform. It becomes a kind of debris centrifuge, flinging off wheels and large chunks of technology. Some photographers duck behind our low retaining wall; others keep shooting. I watch the erratic flight of the car and its former car parts, consider my options and run for it. The chassis catches the grass and slews well away from us. Crawford is shaken up but uninjured.

Suddenly I have less nostalgia for the racing cars of my youth. I'm told that Crawford is a very nice fellow. Now he can have dinner tonight and then come back and qualify tomorrow in the spare car. In one of the old roadsters, he would have been killed.

This year's great practice controversy is about the diffuser rule. USAC has tried to slow the cars down by restricting the airflow through the sidepods. The 1990 cars have been designed around the new rules, but the old ones have been modified, some with more success than others. Jeff Wood, Jeff Andretti and Johnny Rutherford have already hit the wall. Bobby Rahal says all the cars have less downforce and are a little trickier to drive, so allowances have to be made. As usual, the teams who did the most wind-tunnel homework are complaining the least and going the fastest. Still, practice

has made it quite clear that to win here you will need a properly balanced, new-from-the-drawingboard 1990 chassis. Preferably with a Chevy engine. The old cars are in a separate race.

Saturday. Pole Day. I wake up in the motel room and it's raining hard enough to make cars pull off the road. A mean, gusty day. My room is in Terre Haute, 75 miles from the Speedway, at a motel my California travel agent described as "just west of the track." I drive toward Indy for an hour through the rain, making a mental note to send my travel agent a map of Indiana. Or possibly have his knees broken by someone named Bruno.

The infield is wet, muddy and dark; garage doors are closed. In the afternoon the rain stops and heroic efforts are made to dry the track with the deafening turbine track dryers and a few thousand laps with service vehicles. The cars come out to pit row, the drivers suit up, but it's no dice. Too wet. The crowd moans and the cars are put away.

Sunday looks like a repeat, but then lightens up and the crews get the track dry for a short qualifying session at the end of the day. Fittipaldi sets fastest time with a 225.301 average for his 4 laps and is joined, provisionally, on the front row by Mears and Rahal. Three Andrettis fill the second row, in a sort of phalanx of genetic consistency. Time runs out, the first weekend of qualifying is over, and Little Al and other hopefuls will have to wait until the next weekend.

When the weekend comes, Emmo's pole position will stand, but others will shift around. The front row will be Fittipaldi, Mears and this Dutch guy, Arie Luyendyk. Row 2 will contain Rahal, Michael Andretti and Mario. Third row, Al Junior, who couldn't find that speed again, then Foyt and Danny Sullivan.

On the way to the Indianapolis airport, I turn on the car radio. The oldies station plays "I Want to Get to Know You," by Spanky and Our Gang and the DJ says it was popular the year Bobby Unser won his first Indy. He doesn't mention the year, 1968, but he probably doesn't have to. This town really gets behind its 500.

Memorial Day weekend I drive from Wisconsin to Indy rather than

fly, to see what the approach feels like from the ground. Across the green hills of southern Wisconsin, through suburban Chicago, where endless tollbooths buy you the worst roads in the nation, then, gratefully, into the prairies of Indiana. On I-65 you begin to see recognizable race fans—motorhomes and vans with Indy decals, Corvettes with radar detectors, smoking Pinto wagons loaded with a beer cooler and four or five guys in cool neon sunglasses.

Finally you see Speedway signs, then turn off at exit 16A and head down 16th for a slow lap around the revelry and madness bordering the Speedway—a gauntlet of tents, vans, rock 'n' roll, Frisbees, souvenir stands, sausage smoke, beer kegs and handmade signs advising young women to take their clothes off. Not many do, but hope seems to be running high.

Only blocks away is a quiet, tree-lined neighborhood where, by word of mouth, I've found a room to rent in a private home. No more Terre Haute. A very nice lady owns the house and rents beds to a friendly band of perennial race fans. You can walk to the track from here. Which I do, at 5:00 a.m.

It's a beautiful spring morning, light haze in the air, first light falling on the maple trees and the silent ruins of last night's parties. Then, when I'm a few blocks from the track, someone cranks up Bob Seger on the van speakers. A collective howl of race-day ecstasy and hangover agony rises from the campground and the morning has started. Sirens, motorcades, honking and traffic backed up on 16th Street. Show your ticket, under the track and into the infield.

In Gasoline Alley, garage doors open, mechanics polish wheels and organize tool carts. The crew for Dean Hall, Indy rookie and sometime skier, practices tire changes while Hall watches. He's up early, the only driver I've seen. Can a rookie sleep before the 500? Can any driver? Hall looks rested and alert.

Indy's stands fill, a photographer friend points out, like the hands of a clock moving. If you watch, you can't see the movement, but each time you look back there's been a change. He's right. By the time the Purdue band marches, the Speedway is a forest of humanity so thick and vast that the overlapping tiers of grandstands acquire the blueness of smoke.

More marching bands, baton twirlers in gold swimming suits, prayers and patriotic songs. The vice president tours the track. Jim Nabors sings "Back Home in Indiana." This can seem interminable on TV or radio, but on the grid, time flies. Drivers put on hoods and helmets, fiddle with belts, adjust mirrors, nod at well-wishers.

Then, suddenly, the thumbs-up smiling part is over. Eyes behind visors disconnect from families and crews, and begin to stare down the track. It's a palpable mood shift that separates the drivers from everyone else, as certain as a slamming door. The eyes, if they make any contact at all, say, "I'm in the car and you're not."

And Mrs. Hulman says, "Gentlemen, start your engines."

I'm standing next to Fittipaldi's car as the engine explodes to life and joins the fantastic mechanical shrieking and howling on the grid. Thirty-three mad dogs, wide awake and barking. Dust rises in mini-cyclones and heat waves wrinkle the air. Time for this kid to clear out and head down to the corner.

The start is thrilling but uneventful, cars pouring into the first turn and sorting themselves out. Emmo is in the lead and will stay there for a long time as Rahal moves into 2nd and Mears immediately begins dropping back with handling problems. Mario, Al Jr. and Michael Andretti are all in the hunt, and so is Luyendyk. Foyt runs 8th, gradually to move up to 6th.

On lap 20 Danny Sullivan wallops the outside of Turn 1 and leaves on the clean white wall a speedy cartoon of two black tires in motion, but there's nothing comical about it. He slides down the track and comes to rest on the outside. Sullivan climbs out of the wrinkled car and signals he's all right. We later learn he felt a

vibration and began to slow when the right rear-wheel bearing came unglued, folding a leg out from under him. Only Emmo is now carrying the Penske torch with any conviction, still in 1st.

But this, too, will pass. After leading for 117 laps, Fittipaldi comes in under the green for new tires. The old right rear is blistered. Pretty soon other drivers have blistering problems as well. And handling problems, and more tire problems. Pit row begins to look like a recently discovered shortcut. After the race, some drivers will complain about their Goodyears' blistering, and Goodyear will immediately issue a statement pointing out that the teams who followed Goodyear's pressure recommendations had no tire problems. Only those who over-inflated their right rear tires in search of slightly looser handling had trouble, increased pressure causing the tread to crown in the center.

A track washed clean by recent rains, along with warmer race-day temperatures, has also added to the heat buildup in right rear tires, and the whole combination seems to have caught most teams by surprise. They keep changing tires and blistering them.

As others mess with tires and handling, Rahal takes the lead on lap 135 and then extends it to almost 15 seconds. The race appears to belong to Rahal, but I casually notice that Luyendyk is really not very far behind. In fact, he's been hanging around with the fast crowd of really famous guys for most of the race, kind of quietly.

Suddenly he's not so quiet any more. It's as if someone has opened all the vents on Luyendyk's Weber grill and the coals are starting to glow white hot instead of dull red. He's been smoldering for the whole race, and now it's time to apply some real heat. On lap 168 he's right there, behind Rahal, and he pulls off a smooth, gutsy pass on the inside of Turn 3. Both drivers pit for tires one last time, but Luyendyk keeps the lead and loses Rahal in traffic.

It's all over. If pistons, valves,

■ **Danny Sullivan had a less-than-perfect Indiana homecoming, thrown into the wall early in the race by an exploding right rear wheel bearing.**

wheel bearings, tires and brain cells keep working right, Luyendyk will have his intelligent, slightly bemused expression cast in sterling silver on the base of the Borg-Warner trophy. He'll be able to say, "I won the Indy 500," but he won't have to. Others will say it for him, for the rest of his life.

He makes no mistakes and brings it home. The Douglas Shierson team's 1990 Domino's Pizza "Hot One" Lola-Chevy—with, appropriately, partial sponsorship from Dutch Boy Paints—flashes under the checker and the pandemonium of well-earned joy breaks out in the Shierson pits. Mieke Luyendyk joins her husband on the victory stand and the crowd takes them in. A smart, good-looking couple. The grandstands begin to chant "Ar-EEE, Ar-EEE," and we've got ourselves another American hero. From the Netherlands, by way of Brookfield, Wisconsin, out of Scottsdale, Arizona, through years of Super Vees and 75 previous Indy-car starts. In all those 75 races, this is his first victory.

It's been a strange race, full of surprises, unexpected failures and undeveloped conflict, where, once again, the winning team is the team that has thought of everything—absolutely everything—and then had good luck on top of it all. It's also been a satisfying race, however. In Shierson and Luyendyk we have tenacity and skill rewarded at last, and it feels good. Glory, it seems, pays closer attention to detail than the rest of us, shuns the obvious and takes care of its own.

When I finally leave the Speedway to drive home, hours after the race, the front grandstands are still far from empty. No "I'm outta here" syndrome for this crowd. They are quietly watching mechanics put the cars away and applauding when drivers emerge from garages or press conferences to walk through Gasoline Alley. Indy folds its tent slowly and reflectively compared with other races I've seen, a little sad that it's all over for another year and probably a little tired. People remain scattered through the stands, alone or in small groups, watching over the empty track, even though it's almost dark.

No one has told them it's just another race.

NEW ENGLAND & SEVEN

Exploring the colonies in a car from those wonderful folks who brought you Bunker Hill

BY PETER EGAN
ILLUSTRATION BY DENNIS BROWN

MY FIRST INKLING we would See New England in Autumn came last year when my wife Barbara returned from the library with an armload of travel books and left them lying all over the house. I noticed they had titles like *New England's 100 Most Expensive and Romantic Inns* and *Sotheby's Guide to Maine Lobster Prices* and *Bed & Breakfast for About the Cost of a New Air Compressor.*

A mild sense of panic set in as I was only half finished assembling an MGB cylinder head, and, as they say, you can't take it with you. It's hard to install bronze valve guides and do a light cleanup of the intake ports at the average bed & breakfast place, as most of them won't let you run a high-speed grinder after sundown. But sometimes a man has to lay down his tools and practice the fine art of relaxation. Every 10 or 15 years seems about right.

So I agreed we should fly to New England (which we'd never seen before), rent a car and take in the fall colors. To save the trip from complete normalcy, however, I called my old friend Chris Tchorzniki and asked if we could visit his car shop while in the Boston area. Chris is a Lotus buff who imports Caterham Sevens, owns a business called Sevens & Elans and collects videotaped episodes of *The Prisoner.* Nearly a perfect human, by my standards.

"Not only can you stop by," Chris said, "but I have a Caterham Seven demo car you can borrow for your trip. While you're in town, you can stay with Mari and me."

Suddenly the vacation seemed like a fine idea. I canceled our rental-car reservations (not that I don't like a good subcompact sedan with a sluggish automatic transmission and 13 x 3½-in. tires) and Barb and I flew into Boston the first week of October.

We took a cab out to Cambridge, where Chris and Mari have the top floor of a lovely old restored brick townhouse, just a short walk from the Harvard campus. The next morning, we walked over to Chris's shop (Sevens & Elans, 248 Hampshire St., Cambridge, Mass. 02139; [617] 497-7777), which is literally around the corner from their house.

Chris threw open the doors on a beautiful red, polished aluminum Seven, with the 135-bhp 1700 Super Sprint engine (a Ford Kent 1600 big-valve crossflow, bored out, with two Webers and a warmed-up cam) and De Dion rear suspension. It was a long-cockpit model—2 in. longer in the driver's compartment than the original Lotus Seven.

He showed us the ups and downs of the top and side curtains and helped us stuff our single small duffle bag (we'd been warned) behind the seats and under the rollbar. It was a cold, clear autumn day, so we decided to leave the top up. As we were getting ready to leave, Chris's friend John Langermann showed up in his B-20 Lancia.

"John and I will lead you out of Cambridge to Route 2 so you won't get lost," he said. "If you don't mind, we'd like to stop at the restoration shop of J. Geils. It's just out of town, near Ayer."

"*The* J. Geils?" I asked. "Of the J. Geils Band?"

Chris nodded. "He took some of his rock 'n' roll earnings and built a very nice shop near his home. He restores vintage racers and works on all kinds of interesting cars. Also likes bikes. He's got a Ducati like yours."

We drove out of Cambridge, through Harvard, past Longfellow's home, around Walden Pond, skirting Lexington and Concord, and I was introduced to the first great truism of this trip, which is that New England (like Old England) is a place of condensed vitality, and there's hardly a name on the map that doesn't evoke some image from history, art or literature. Especially if your 8th-grade teacher was Miss Dillon, who treated you mean and made you learn stuff.

Near Ayer we pulled into a wooded drive that led to a large garage/workshop where we met J. Geils, whom Barb and I had last seen from about a mile away at the Los Angeles Coliseum, where his band opened for the Rolling Stones. At shorter range, he's a youthful-looking, personable gentleman without a trace of rock-star affectation, a man who likes cars and seems to do excellent work on them.

You have to admire people who can use both sides of their brain. If I played guitar as well as J. Geils, I'd be sitting in a dark room somewhere, staring off into space.

After a tour of the shop and a discussion on the merits of Ducatis, we

■ Below left, an inspiring view—the Seven's hood, the New Hampshire countryside and the imposing Mt. Washington Hotel. Below right, sweater-clad J. Geils and the author.

Humans, we suddenly discovered, have a finite capacity to enter and exit a Caterham Seven and then they run plumb out of energy and seize up. Especially after dinner.

■ Tour-bus travelers took a liking to the Seven.

said goodbye to Mr. Geils and our Lancia escort and headed into the wilds of New England unescorted. Highway 111 to 119 to 124, into New Hampshire through New Ipswich and Jaffrey.

We'd no sooner crossed into New Hampshire than I saw a basic error in our planning of the trip: All our hotel reservations were too close together, separated by no more than three or four hours of driving. One too many trips across Texas had distorted my perspective on maps. New England is a small place, and it doesn't take long to get anywhere. Our first stop for the night was at the Amos A. Parker House in Fitzwilliam, only a few hours' drive from Boston.

So to kill time we jogged, detoured, looped and backtracked around the rural New Hampshire countryside, which led to a couple of other discoveries. First, the roads in New England are wonderful. They go up and down hills, through dark woods, descend into mossy glens and emerge into sunlit hilltops of fiery red and orange maples. The roads curve and crest endlessly and there's not much traffic, even in the tourists' favorite season. What traffic there is doesn't bother you much because the scenery is so beautiful you don't mind cruising and looking around.

Second, the Caterham Seven seems to have been tailor-made for these roads. At New England speeds the Seven is just prowling most of the time, muttering to itself and waiting to get up and howl around the occasional delivery truck or slow sedan. It has so much reserve in cornering power and acceleration that it is limited only by the driver's ability to see (and think) ahead. It corners dead flat, soaks up road bumps nicely and goes like hell when you put your foot in it. There are no mechanical limitations, only those of conscience and public safety.

In that respect, it's an almost perfect tool for exploring back roads. Except that it has a lot more charisma than a mere tool, and it attracts more attention as we discovered when we stopped at a roadside war memorial called the Cathedral of the Pines.

We pulled up next to a busload of elderly tourists and an old gent shouted, "You want a can opener, so you can get out of that thing?"

Cackling and laughter came from the tour group, which soon clustered around the Seven, much to the consternation of the tour director, who was trying to herd them toward the Cathedral of the Pines. "Is that an MG?" someone asked. We spent about half an hour explaining the car while storm clouds gathered in the tour director's face. Finally, we wandered off to look at the memorial park and chapel, which turned out to be a touching monument to a young aviator killed in World War II. It was built by his parents.

When we came back to the parking lot, many in the group had ignored the Cathedral and were still examining and discussing the Seven. Barb and I jackknifed ourselves through the small side openings, slid into the seats and waved goodbye.

When I looked into the mirror, all those gray heads were still watching us drive away, hands waving. The Seven seemed to remind older people of something they missed—or perhaps had missed out on. Maybe it suggested a Model A with a few layers missing, an underslung roadster that couldn't be rolled. Or freedom from tour directors.

We checked into the charming old Amos A. Parker House, an erstwhile meeting spot for Revolutionary War troublemakers. Its owner, Freda Haupt, once lived near my hometown in Wisconsin, so we had a good time comparing notes. Barb and I later had dinner at a very old restaurant, where the food seemed to have been prepared by a person trained for some other profession, possibly carpentry or nuclear physics.

In the morning we weaved westward on Highway 119, across the Connecticut River into Vermont, then Highway 9 to 100, north along the edge of the Green Mountains; another day in sports-car paradise. As we passed through more and more small New England towns, I realized that most of the smaller villages seemed to exist completely on tourism, a reflection of the sparsity of farms and agriculture. Seems most New England farm boys came home from the Civil War just long enough to pack up and move to the prairie states, where you could "plow all day without hitting a rock." As a result, the dense woods

of Vermont and New Hampshire are laced with ancient stone walls that once marked the borders of pastures and fields, now overgrown with trees. A little empty and lonely, but great for driving.

Perhaps these departing farmers should have left the Mohicans a quick note: "Thanks for the neat woods—you can have 'em back," or words to that effect. An agricultural forerunner to the test drive.

Crossing over the ridge of the Green Mountains, we took aim on Big Equinox Mountain (3816 ft.) and paid $6 to drive up the private road where the Equinox hillclimb used to be staged. A spectacular drive with a weather-beaten hotel and restaurant at the windy top. We descended for the night into Manchester, where Revolutionary charm blends unhappily with factory outlet warehouses that sell clothing designed to make New Yorkers look like sheep ranchers from the Isle of Skye, which is no easy trick.

We pulled into the long driveway of the beautiful old Inn at Manchester, then had an excellent dinner at Ye Old Tavern, built in 1790. When we left the restaurant and walked back to the Caterham, Barb and I both stopped dead in our tracks and stared at the car. We'd had the top up all day and had already climbed in and out of the door/window openings perhaps two dozen times. Humans, we suddenly discovered, have a finite capacity to enter and exit a Caterham Seven and then they run plumb out of energy and seize up. Especially after dinner.

"We could walk back to the hotel," Barb suggested.

"No," I said, quoting a former president. "That would be wrong."

We climbed into the Seven one more time, upholding an ancient Lotus code of honor, a tradition that hinges on the Zelda and F. Scott Fitzgerald pledge never to admit you're too tired to do anything.

Awaking to a dark, cloudy day, we cruised past Hildene, Robert Todd Lincoln's home, and crossed the mountains to Highway 100 A, stopping to see the country homestead of Calvin Coolidge, which is a fascinating museum to this modest, thoughtful man. The Coolidge farm is like a monument to a bygone era of Yan-

kee frugality, well worth a stop for those burned out on our present heroes of the Savings & Loans Era.

Highways 118 and 112 took us through the White Mountain National Forest, about 50 miles of the finest sports-car roads encountered in my short life, and then we hit the first truly crowded highway of the trip, Highway 16 North on a Friday afternoon. In rain and growing darkness we drove up to the famous Mt. Washington Hotel, a sprawling, beautiful old railroad resort built in 1902. From a distance, it looks like the hotel where *The Shining* was filmed, though Jack Nicholson and his fire ax seemed mercifully absent.

Instead, there are broad porches and a spectacular lobby, dining room and ballroom. I needed a jacket for dinner and, of course, didn't have one, so the bell captain directed me toward a whole closet of loaners for just such an occasion. A fine collection of Soupy Sales plaids and loud houndstooth checks in cranberry and gray polyester, presumably the legacy of guests who were quietly murdered by the maitre d' before they could ruin the tasteful elegance of his grand dining room.

I finally found a tan cotton sports coat, and I told Barb that, with my matching khaki slacks, I felt like Denys Finch-Hatten at the Nairobi Hunt Club. She said I looked like a vagrant who had stolen a jacket from a dwarf. Nevertheless, we danced the night away to an excellent Glenn Miller-style band in the grand ballroom, and no one threw me out.

When we left in the morning, the elevator operator, a small woman in a red bolero jacket, said of the weather, "Cold out they-ah. Not used to it, you'll soon be toe-dancing."

We toe-danced to the Seven, turned on the surprisingly effective heater and headed snugly toward Augusta, Maine for a reunion with my aunt and uncle, Margaret and Charles Schneider, and all my cousins. They have a summer cabin, which in this part of the country, we were told, is called a "camp."

After a wonderful weekend visit, we drove down the coast to Boothbay Harbor and then Kennebunkport. The latter is a lovely old sea coast village whose downtown has become uncomfortably touristy, full of the usual T-shirt shops, "olde" fudge stores and "art" galleries. President Bush was visiting his Kennebunkport home while we were there, so we drove along the coast to the edge of the compound and waved, but were not invited in.

From there we inched southward through a succession of weekend traffic jams, back toward Boston. Approaching the city from the north, we were carried along on the shoulders of the traffic, deflected toward Cambridge and bounced via pothole toward the driveway of Sevens & Elans. Home.

We had only two days to explore Boston, a great, slightly tough city with chaotic traffic, wonderful oysters (at the Union Oyster House), superb Italian food on the North End, good local beer and enough historical magnificence to keep Miss Dillon busy for a decade. She would be proud to know we walked the Freedom Trail, toured Harvard and Longfellow's home and walked all the way around Walden Pond before we turned the Seven back over to Chris and flew home.

If we were to do it all again, I think we'd dedicate more time to the exploration of Boston and then spend all our driving time on the farthest back of back roads, seeing more of northern Maine. Perhaps. The same car would do just fine. We'd also watch more of Chris and Mari's *Prisoner* episodes.

Purely from a sports-car driver's point of view, the interior roads of New England seem far more pleasant than the Atlantic shoreline, with its concentrated tourism. I say this from the vast experience of one autumn weekend on the coast. Maybe it's better during the week. But there's such an incomparable freedom and remote beauty to rural roads of Vermont, New Hampshire and Maine that there seems little reason to gravitate toward the population centers and their crowded arteries. In the Green and White Mountains, you can still drive through country rather than real estate.

Add an explosion of autumn color and a good sports car, and you have an excellent case for setting aside the air grinder, turning off the garage lights and leaving your cylinder head behind. ◉

HOVER DYNAMICS RX2000

Blender in the grass—not to mention the swamps, beaches, drag strips, oceans and parking lots

BY PETER EGAN

PHOTOS BY DEAN SIRACUSA

EVITATION, THAT'S THE original dream. Not free, soaring flight, but simple freedom from the ground. An invisible force that holds us just off the surface of the earth and makes travel a smooth, beaming glide over rocks and potholes.

Luke Skywalker had the right idea, with his beat-up hot rod of the future that hovered over its parking spot or hummed along at high speed, just a foot or two above the trackless desert. Magicians from time immemorial have been amazing us by levitating scantily clad women above tables and passing hoops around them. Look, no strings. How do they do it? (Notice that hardly anybody ever says, "Hey, what's the point?") A sorcerer's first proof of power is to point at some hapless object—or person—and make it hover off the ground.

Hard to say why we like levitation so much. Revenge on gravity, perhaps, the evil force that breaks our drink glasses, steals our baseballs and skis, gives us double chins and back trouble and even kills us when we drive off a precipice at Big Sur or do Fred Astaire impressions on the penthouse balcony railing. After only a few years on earth, most of us have had it with gravity. At the first glint of childhood imagination, we're ready to levitate.

Unfortunately, we don't have any money. Much later, however, we go to school, learn a trade, work hard, buy a house, replace all our records with CDs, buy espresso machines and discover we still aren't broke. We're older and wiser, but the desire to levitate remains and we have a credit line worth $9000. We're ready for the RX2000 hovercraft.

Okay, but where do you get one? Well, you call up the folks at Hover Dynamics in Cumming, Georgia and ask nicely if they'd bring one all the way to California for testing. They say sure, and two guys, Ken Morgan (test driver) and Gary Woodruff (product engineer), hook a trailer onto a pickup and haul one through fog and sleet and Texas snowstorms and show up at your office on a Tuesday morning. They start the hovercraft engine and the RX2000 literally floats off the trailer, buoyant as a watermelon in a horse trough. (Now there's a simile you don't hear around the shopping mall.) It's 520 lb. of dead weight, suddenly weightless.

Look at the thing: It's 13.1 feet long, a finned fiberglass boat shaped like the Space Shuttle. But it's no ordinary boat. There are holes around the edge of the hollow hull and 66 skirts of neoprene-impregnated nylon to seal off the cushion of air un-der the hull. The air comes from a Rotax 52-bhp 2-cylinder 2-stroke engine with a 9-blade propeller. The shrouded prop blows air rearward, where a shelf divides the airflow, directing 60 percent of it into the hollow hull and out those bottom holes, while the other 40 percent blasts rearward through twin rudders, providing control and forward thrust.

Controls are simple: a steering wheel and a throttle lever. There's also a tach and temperature gauge on the panel, and you sit behind it, straddling the snowmobile-type bench seat, which has room for two.

Before the engine is started, the RX2000 rests on three long aluminum skids. No parking brake needed; it doesn't want to go anywhere, and won't. We climb in, flip on the choke, hit the electric starter and the Rotax comes to life with earsplitting urgency and enough wind to blow dust down your neck and fallen leaves into the next county. As revs climb, the skirts billow and close ranks, the hull puffs out slightly and it reminds you of a large opera soprano taking a deep breath. More revs and the song begins. The craft rises majestically off the ground, bobbing and floating. At this stage, someone standing on solid ground can push it around like a pallet on billiard balls.

Bring the engine up to about 5000 rpm, however, and the RX2000 starts to waft forward and pick up speed. Ken gives us driving lessons in the parking lot. Turning requires a combination of rudder and weight shift, almost like an ultralight aircraft, along with forceful blasts from the propeller. "Don't chop the power suddenly, especially in a turn," Ken says. "Always ease the power off."

I put the cord for the "dead man's switch" around my wrist (it turns the ignition off if you fall out) and try the craft myself. It's eerie at first. A strange mixture of terminal understeer and oversteer, it wants to follow the gravitational crown of the road and fall off toward the curb, unless you crab away from it with the power on. Turns are initiated 20 or 30 yards in advance. Control is anticipatory and steering has all the precise immediacy of a Christmas wish, or some other vague longing. Not since the VW microbus have crosswinds been able to effect such an effortless lane change on a vehicle.

Practice does not make perfect, but it helps a lot and you gradually get so you can place the hovercraft

The RX2000 is not at its best in point-and-shoot traffic. There's a floaty sensation during acceleration and cornering, and brakes can be grabby.

just about where you want it in the large parking lot next to the Pomona drag strip. The RX accelerates across the pavement to speeds of around 45 mph, and when the fences start looming at the end of the asphalt, there are two basic ways to turn around. The simplest is to cut the power and ease to a stop on the skids, then hit the throttle again and do a stationary 180 with the rudders.

The more daring, stylish approach is to do what I call a Wayne Gretzky. This is a flying 180 where you pivot the hovercraft on its axis so that you fly *backward* toward the fence like an ice skater at high speed, then give it full throttle and hope it slows down and reverses direction before you run out of pavement. In truth, vehicle dynamics are closer to those of a hockey puck than they are to the sharp control of an ice skater. Road Test Editor Kim Reynolds points

out that these reverse 180s make you feel exactly as if you're on the end of a bungee cord that has just been stretched to the limit and is starting to contract.

Time for real testing and some hard numbers. With engineer Gary Woodruff aboard, the RX2000 picks up its skirts and scurries down the quarter-mile strip in 33.0 seconds at 44.0 mph. There's an unearthly, miraculous look about its ac-

celeration that reminds me of Jake and Elwood's Mother Superior in the *Blues Brothers*. You keep looking for feet or roller skates underneath the habit but there aren't any. Joe Blackstock, Hover Dynamics' West Coast representative, tries his luck and gets a similar time. He returns to the starting line, tries a slightly late Wayne Gretzky Turn and clouts the Armco, putting some stress cracks in the fiberglass.

Slalom cones next. With a little practice, Gary and Ken hustle the craft through the slalom in just 30 sec., but on one run Gary turns wide and hits the Armco again. More minor glass damage.

"You wouldn't want to drive one of these around in your living room," Kim Reynolds observes.

He has a point. Especially if you collect china figurines or Ming vases. The RX2000 is not really happy maneuvering in a restricted environment full of hard objects. Like the personal helicopter, it is not the undiscovered answer to America's commuting woes. It's obviously a craft intended for the wide-open playground—beaches, open water, sand dunes, ice-covered lakes. The owners have been more than patient in letting us test it in an automotive setting. The next day we head for the ocean.

Now we're talking. The Pacific Ocean, we discover, is relatively level and has almost no Armco. Even in Newport Harbor, there's more room to maneuver. On our test day, the ocean is smooth with only light swells and the RX2000 is able to get up to 45 mph, wide open, and stay

With a deft combination of weight transfer, steering input and facial contortion, Gary coaxes the craft through our slalom. Two-stroke, 2-cylinder Rotax engine makes 52 bhp.

HOVER DYNAMICS RX2000

PRICE

List price, FOB factory **$7995**
Price as tested **$8980**

  Price as tested includes std equip. (bilge pump, port & starboard running lights, strobe light), 52-bhp Rotax engine ($900), dash-mounted fuel level gauge ($85).

MANUFACTURER

Hover Dynamics Manufacturing, Inc., Rt. 12 Box 774F, Cumming, Ga. 30130

0–40 mph	25.0 sec
0–¼ mi	33.0 sec
Top speed	46 mph
Skidpad	0.09g
Slalom	14.5 mph
Brake rating	not bad for wearing a skirt

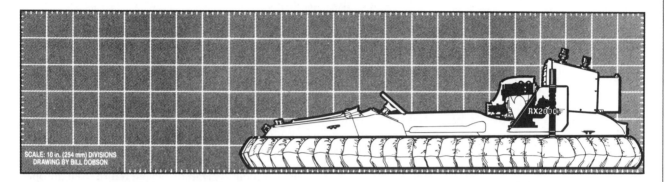

SCALE: 10 in. (254 mm) DIVISIONS
DRAWING BY BILL DOBSON

ENGINE

Type .... aluminum crankcase, cylinder & head, **inline-2, 2-stroke**
Valvetrain .... none (cylinder ports)
Displacement ..... 30.3 cu in./497 cc
Bore x stroke .... 2.84 x 2.40 in./ 72.0 x 61.0 mm
Compression ratio 10.8:1
Horsepower
  (SAE) .... **52 bhp @ 6600 rpm**
Bhp/liter 104.6
Maximum engine speed .... 7000 rpm
Fuel delivery .... 2 Bing carbs
Fuel requirement .... 50:1 gasoline/ oil mix

CHASSIS & BODY

Layout .... **rear engine/rear drive/ underside levitation**
Body/frame ..... fiberglass/reinforced fiberglass
Brakes
  Primary .... **skid-on-ground type**
  Assist type... 180° spin, full power
Skirts .... 66, urethane coated nylon, pleated in basic blue
Steering type .... **twin rudders**
  Assist .... push-pull cable
  Turns, lock to lock 0.5
  Turning circle .. on a dime or better
Suspension .... compressed air

DRIVETRAIN

Transmission ....**propeller**
  Blades 9
  Diameter .... 24 in.
Final drive ratio 2.0:1

GENERAL DATA

Curb weight .... **520 lb**
Test weight .... 0 lb
Weight balance (with driver),
  % length from bow 42
Hoverbase .... 12.5 ft.
Hovertrack .... 6.7 ft.
Length .... **13.1 ft**
Width .... **7.3 ft**
Height .... **45.0 in.**
Ground clearance .... 0–8.0 in.
Trunk space .... 1.6 cu ft

FUEL ECONOMY

Normal hovering .... 2 gal./hr
Hovering time .... 3.0 hrs
Fuel capacity .... 6.0 gal.

ACCOMMODATIONS

Seating capacity .... **1+1**
Head room ..... plenty for standing or even jumping
Seat width .... 11.0 in.
Leg room .... 54.0 in.
Seatback adjustment .... propeller limited

INSTRUMENTATION

8000-rpm tach, cylinder head temp, fuel level

MAINTENANCE

Air/fuel filter inspection .. 10 hrs/10 hrs
Cylinder decarbonizing .... 50 hrs
Basic warranty .... 6 mo

ACCELERATION

Time to speed	Seconds
0–10 mph	3.0
0–15 mph	6.0
0–20 mph	9.0
0–25 mph	13.0
0–30 mph	15.0
0–35 mph	20.0
0–40 mph	25.0

Time to distance
  0–1320 ft (¼ mi): 33.0 @ 44.0 mph
Top speed .... 46 mph

BRAKING

Minimum stopping distance
  From 30 mph (on asphalt) ... 117 ft
Brake feel .... fine on water, but asphalt is grating
Overall brake rating .... not bad for wearing a skirt

HANDLING

Lateral accel (200-ft skidpad) .. 0.09g
  Balance .... variable oversteer, sometimes even backward
Speed thru 700-ft slalom .. 14.5 mph
  Balance .... tippy if you stand

INTERIOR NOISE

Idle in neutral .... 57 dBA
Full throttle @ 50 ft .... 78 dBA

Subjective ratings consist of excellent, very good, good, average, poor.

Test Notes . . .

■ The RX2000 rides as if it were on a cushion of air, which, of course, it is. That's fortunate, because what it rides over is unpredictable. Often it responds to steering inputs. Sometimes not.

■ Braking from low speeds amounts to cutting the throttle and skidding it home. But from top speed it's better to spin the RX2000 backward and stand on the gas. You stop sooner and don't see what you're heading for.

■ Grip off the line is poor, but once underway, the RX2000 can really pick up its skirts and kick sand in your face. That's in addition to discarded pennies and cigarette butts.

there for miles. It's marvelous. I can't think of any other 52-bhp craft that goes this fast over the water, and without a hint of vibration or hull slap. The ride is glassy-smooth.

Jumping light waves proves to be good fun, and I ask Gary if heavy

Interior is Spartan, even more so than a BMW's. Below, airbag is inflated whenever the RX2000 is in motion; a coup in safety engineering!

seas are a problem. "I tell people not to take it out if the waves are too rough for a regular 13-ft. motorboat," he says. The RX2000 corners flat, with almost no listing of the hull, so I can imagine that sliding sideways into a real wave could be, well, wet.

Speaking of wet, the hovercraft covers you with a light spray during acceleration and fast turns, so your clothing gradually soaks up plenty of sea water. It's not exactly a wet boat, like a Sunfish, but wearing an expensive worsted wool business suit from the Botany 500 Johnny Carson collection would be a big mistake.

While fun on both water and land, the true miracle of hovering does not make itself known until we run the RX2000 up on beaches and boat landings and back into the water. You head for the shore at full speed and every instinct of self preservation screams at you to chop the throttle or jump out. But the hovercraft hits the shore and wafts up onto the beach without so much as a bump to tell you the medium beneath the hull has changed. In fact, cutting the throttle while entering or leaving the water is the worst thing you could do—automatic ejection city.

Sandbars, reefs, spits and marshes all glide beneath the RX2000's hull without comment. So do animal habitat and bird nesting grounds. In that respect, the hovercraft has the

potential to be an environmentalist's nightmare. It allows man to go where no man has gone before, which is not always a good thing. On the other hand, it *can* go into sensitive habitats without touching the ground, chopping weeds, muddying the waters, lacerating manatees, breaking eggs or leaving a mark on the earth, so it has the potential to be a useful tool. In the hands of a dimwit yahoo, the RX2000 can be just another noisy, destructive toy.

In the hands of my hometown fire/rescue unit, however, which just bought one, it's a means—maybe the only means—for rescuing ice fishermen, cross-country skiers and hikers who have fallen through thin ice. A hovercraft depends a lot on its driver for image.

It also depends upon the Harbor Patrol for passage through Newport Harbor. On our return from the open ocean, a patrol boat stopped us and said hovercraft are illegal anywhere in the county. We'd have to shut down the fan and tow it back to the landing with our camera boat. Why?

"Unsafe. Too hard to control in harbor traffic."

I wasn't having any trouble, but perhaps a dimwit yahoo had preceded us. The officer didn't say "too loud," which surprised me. The RX2000 is terrifically noisy, requiring earplugs for the driver and garnering its share of dark looks from boaters. Fortunately, Hover Dynamics has a quieter, more efficient fan on the way later this spring, along with a quieter, 4-stroke engine. In the meantime, those who are rescued from drowning on half-frozen lakes or in boating accidents probably won't object to either the noise or the speed of the RX2000. Nor will they mind its ability to take shortcuts over land. They'll also be grateful it doesn't have a sharp metal prop spinning in the water.

Equal parts boat, airplane, car, leaf-blower, hockey puck, helicopter, vacuum cleaner, snowmobile, margarita blender, Jet Ski, bumper car and floor buffer, the RX2000 provides advantages and disadvantages taken from each. Under the right circumstances, nothing else will do, and the hovercraft is pure sorcery. In suspect terrain, levitation is the only way to go. ◉

GONE SOUTH

A visit to Charlotte, North Carolina, home of good times, fearless drivers and the world's fastest sedan-shaped objects

BY PETER EGAN

SOMEWHERE ON HIGHWAY 421 in central Kentucky we began to see signs of the South. Not the modern, urban South that architecture and prosperity have made indistinguishable from cities anywhere else in the U.S., but the mythical, rural South. The one we like to believe in, the one where stock car drivers are supposed to be made:

Hand-lettered signs pertaining to Jesus and personal salvation, shotgun shacks on cement blocks with dogs of uncertain breeding sleeping under them, homes converted to part-time Pentecostal churches, red clay roads disappearing into valleys thick with green hardwoods, signs for Mail Pouch Tobacco and Goody's Headache Powders. On the radio, Reba McEntire and Country FM 102 say hello to Charlotte Motor Speedway. See you there.

Good combination, Reba and Charlotte. Stock car racing has always reminded me of Country music. It has thousands of hardcore fans who grew up with it and love it, while others remain mystified by its appeal. Still others grew up with it but left it behind for something ostensibly more urbane, only to drift back.

And then there are the converts, maybe the greatest enthusiasts of all. They've happened onto a Good Thing and know it, the way you discover Cajun food or Doc Watson even though you grew up in North Dakota and your parents raised you on macaroni and cheese and Lawrence Welk.

Among the converted is an ex-Long Islander named Tom Cotter. He grew up on road racing at Bridgehampton and Lime Rock, and, even now, his garage is filled with sports cars, including

a Lotus Elan that used to belong to Graham Hill. He moved South, however, and fell in love with the world of NASCAR racing, gradually to become public relations director for Charlotte Motor Speedway. Now he works for a couple of racing teams and their sponsors, Country Time Drink Mix and Maxwell House coffee.

Through it all, he's been urging me to come down and see a race at Charlotte, much the way any good tent preacher would call on wayward members of the flock to get up out of their seats, come forward and bear witness. "Come on down," he said, "and see a different kind of racing. The cars turn left, the racing is close, and I guarantee you'll have a good time."

Not much pleading was needed, in my case. Only an opening on the calendar. Having grown up around small town stock car racing, I've always wanted to see a big-time NASCAR race on one of the fast superspeedways. And the Coca-Cola 600 at Charlotte is about as big time as they come. It's an event that works hard at pulling the Memorial Day crowd away from Indy, which, of course, is a mere 500-miler.

So at the end of May, my friend/co-driver Pat Donnelly and I headed out, driving R&T's long-term Ford Explorer southeast from Wisconsin. We picked up Highway 421 in southern Indiana and took this winding, greenly mountainous path all the way into

North Carolina, across the Blue Ridge crest, descending into Charlotte on a Wednesday afternoon.

Charlotte is a usefully sized city with a sparkling, revitalized downtown and a real skyline of steel and glass. It has a large central business district where the streets grow very empty at night, almost like Wall Street, everyone gone home to the suburbs. Or across the tracks to the Longhorn Steak House, a live-wire racing hangout and renowned steak place. We checked into our downtown hotel and checked out the Longhorn. Good steaks. The next day we drove out to the track. Thursday.

Charlotte Motor Speedway is 12 miles north of town in the rolling hills east of I-85 and looks, on approach, like an extra-large baseball stadium. Campgrounds near the front gate are lined with souvenir stands, semitrailer replicas of team transporters, covered with sponsor logos that read like a shopping list for a person with a big thirst, a well-maintained car, a chewing habit and a clean house: Miller, Budweiser, Maxwell House, Folgers, Country Time, Pennzoil, Quaker State, Skoal, Levi-Garrett, Tide, etc. Big Business.

A tunnel takes us under Turn 4 to the infield, crowded with the same trucks and colors, but with race cars and teams living in them. The transporters are works of art, rolling air-conditioned refuges from pit life, with kitchens and lounges. When a door opens, it's like a glimpse into a cool Spanish courtyard from the sunbaked street outside.

The working paddock is a little more down-home. No Formula 1 garages full of computer technicians here. The cars are lined up, in order of championship points, under an open, tin-roofed structure, crowded cheek-by-jowl with mechanics, floor jacks and engine hoists.

And right over there, leaning on the Number 43 STP/Pontiac, is none other than Richard Petty in his famous cowboy hat and sunglasses. His cowboy boots

■ The King, Richard Petty, presumably making some last-minute adjustment to No. 43.

appear to be made from the hide of something that can kill a water buffalo with a single bite, and he looks as thin and lanky as Hank Williams. In fact, Richard Petty probably is the Hank Williams of stock car racing, though considerably healthier than Hank ever was.

People see him, stop in their tracks, take pictures, smile and say hello. He smiles back and waves, talks to his mechanics as he signs autographs. He is easily as famous as anyone in America. He won his 200th NASCAR race in front of the president in 1984, on the 4th of July.

Nearby, in the shade of the race trailer, Richard's son, Kyle, is sitting in a big stuffed chair, his leg in a cast. He broke his leg in a recent accident at Talladega and won't be racing until later in the season. When Kyle gets up and leaves on his crutches, I see there's a note pinned to the chair. It reads, "This is my chair. Don't sit in it.—Kyle Petty."

■ At left, Kyle Petty reclining; above, his recliner.

Looking at the note, I try to imagine if there is anyone alive who would disregard it. Certainly no one in the paddock. Or North Carolina.

If a senator or a state governor left a similar note, about 49 percent of the population would sit in the chair immediately. There's a definite hierarchy of status here. Kyle is the son of the King. A prince.

Other royalty are also wandering around the paddock and infield: various Waltrips . . . Allisons . . . Wallaces . . . and I'm amazed at how open it all is. Everyone is polite and friendly. Stand and look at a car for a few minutes, and the mechanics or drivers will talk to you, unsolicited. Formula 1 teams could come here to study the use of restrictor plates on the flow of self-importance.

Crews finally push the Winston Cup cars out to pit lane and fire them up for practice. Pat Donnelly and I decide to cross under the track and watch the session from Chicken Bone Alley, the cheap seats next to the fence at the exit of Turn 4. There's a big crowd for practice and qualifying, but no one else is sitting here. There are big gummy chunks of spent Goodyear littering the benches. Maybe the crowd knows something . . .

Dale Earnhardt and Michael Waltrip come through the banking like a pair of F-14s in close formation, head directly for our seats and then decide to stay on the track, sliding right up to the wall in a whomping Doppler explosion of sound that all but blows our eardrums out. *eeeeeeeEEEEEE Bang! Baaaaaaa.* When they're gone and the air has collapsed inward to fill the gap, I look at Pat, half expecting to find crossed band-aids on his face and a raccoon mask of tire dust around his eyes.

Pat shakes his head. The Winston Cup cars are qualifying here at around 174 mph and hitting nearly 200 mph on the straights. Not bad for sedans, or at least sedan-shaped objects. They move a lot of air out of the way.

And they really do thunder. The echo of engines and wind boom precedes them through Turn 4, the concrete bowl of the north banking cupping the sound like a jai alai racket and flinging it at your ears. Nothing but V-8s will ever do for this sport. If Detroit ever stops making big motors, STP or Tide will have to do it, just for NASCAR. New for 1998, the Proctor & Gamble V-8.

When qualifying is over for the Coca-Cola 600, Mark Martin is on the pole with his Folgers/Ford, at 174.820. Michael Waltrip's Pennzoil/Pontiac is just behind him. In the afternoon, we have Grand National qualifying for Saturday's Champion Spark Plug 300. There's also practice for the Sportsman Division, NASCAR's minor league for up-and-comers. Nothing major happening on Friday, so our friend Tom Cotter offers to show us around.

We begin with lunch at the Sandwich Construction Company, a popular race-crowd hangout and restaurant, festooned with damaged body panels from famous Winston Cup cars. The iced tea flows freely and the barbecue sandwiches are terrific. While we eat, Tom explains Winston Cup economics.

At one billion dollars a year, it's the fourth biggest industry in North Carolina, after tobacco, textiles and high tech. About 75 percent of all Winston Cup teams and drivers live within 100 miles of Charlotte. So many drivers and team owners have homes on the shoreline of nearby Lake Norman, in fact, that they've wryly dubbed it "The Redneck Riviera."

How important is the advertising on the cars?

Extremely, he says. Hardcore race fans have instant recognition of driver/car/sponsor combinations, and it sells products. Folgers coffee, for instance, got into racing and immediately began to erode the sales of Maxwell House, which had always been the best-selling coffee in the South. Maxwell House put its name on the side of Junior Johnson's Thunderbirds, and now its sales are up again.

Always highly suggestible, I sit there wondering if I should do a private coffee comparison test when I get home. Maybe it really works . . .

After lunch, we stop nearby at the Bahari' Racing works, home of Michael Waltrip's very fast yellow Pennzoil/Pontiac(s). Here we have further evidence that Winston Cup racing is not done on a shoestring. In their huge, hospital-clean garage complex, the team has no fewer than 10 cars, at about $100,000 per copy, all with Waltrip's name over the door. They have a couple of cars for superspeedways, a pair for road racing courses, a few short-track oval cars, and probably one or two for tracks with air-conditioned restrooms. Nothing is overlooked.

Car builders David Watkins and Paul Charcut show me a new chassis under construction and walk me through the basics. You start with a 1965 Ford station wagon floorpan, as used by Holman & Moody in 1965, and build a tube-frame structure over it. Chrome moly? "No, .095 mild steel. When they hit the wall, these things gotta wad up." Stock windshield with Lexan behind it for safety and support because the stock one "wants to cave in" at speed; stock sheet metal from roof, hood and trunk has to fit the template for that model car; 1960 Ford 9-in. rear end, 60-in. track, 110-in. wheelbase. No rear anti-roll bar.

You've got "front steer" and "rear steer" cars, depending on whether you mount the steering rack in front of the tie rods or behind, a matter of driver preference. The ubiquitous Basset 15-in. wheels carry an inner and outer tire casing, for safety.

This week at Charlotte they're trying tubeless inner tires for the first time, trying to eliminate the wheel imbalance judder of flat inner tubes. Two valve stems per wheel, two different inflation pressures. Also different inflation pressures front to rear, right to left. Eight pressure settings per car. No slow-thinking mechanics need apply.

Engines? Chevy or Ford V-8, depending upon the team, 358 cu. in. max, mated to either a Jerico gearbox for road racing or a G&G box for ovals. Total car weight, 3500 lb., 1600 of it on the right side, about 300 lb. of the weight in movable lead.

Off to the next shop. Tom Cotter makes a phone call and comes back grinning. "How would you like to go to Ingle Hollow and visit with Junior Johnson?"

Junior Johnson, the legend. Moonshiner turned stock car driver; did real prison time for running white lightning; a famous driver made even more famous to the non-racing world by Tom Wolfe's *Last American Hero*. The quintessential stock car driver, now a team owner, a man with the entire history of the sport

etched into the lines on his face.

A rare opportunity, but is this a good time to visit? He might not be in a very good mood. His Budweiser/Thunderbird team has been suspended for four races because the engine in driver Tommy Ellis' car was found to be 3.8 cu. in. oversize after last Sunday's Winston race. To get around the restriction, the car has been entered in the Coca-Cola 600 under the name of Junior's wife, Flossie. Ellis will be allowed to drive, but Junior is banned from the pits. The local papers are full of this story, and a great debate is raging.

Junior claims that the slightly oversize engine was an error, caused by a mechanic using the wrong crank with the right block. Other drivers and teams agree that's probably what happened. No one cheats for 3.8 cubic inches—not enough performance gain. Besides, Junior knew the engine would be torn down, so why do it? The car finished 14th anyway. Big deal.

Yet NASCAR has thrown the book at Junior Johnson, mainstay of Winston Cup racing, national hero and local god, telling him to stay home. He's run at Charlotte every year since the first race, in 1960, and his team has won six of the last 15 Winston Cup championships. So why are the officials coming down so hard? Major mystery.

Some suggest it's because Junior Johnson nudged NASCAR President Bill France Jr.'s pace car off the track in the Winston Legends Race, an old-timer's event held last Sunday. Perhaps Bill Jr. was not amused. Others say it was only a technicality to get Junior for past friction with NASCAR over rules and racing policy.

As I listen to these discussions over meals and drinks, and read the stories in the paper, it gradually becomes clear that NASCAR racing is kind of like Chinatown, as interpreted by the movie of that name: You can observe the surface, but you never really know what's going on underneath. Winston Cup racing is first and foremost a show, and constant adjustments are made to keep the show working. Fairness seems not to be specific, but general and long-term. Rules can be bent or closely observed, depending upon the demands of showmanship.

That's what's odd about this one. Junior Johnson is usually a big part of the show. Everyone is confused.

We drive 75 miles north of Charlotte and suddenly turn off the highway at a Pit & Go General Store, onto Ingle Hollow. Set against the green hills, 700 acres of them, is a cattle ranch and farm

Continued on page 127

GREAT SMOKY, BIG HEALEY

Country roads, take us home

BY PETER EGAN
PHOTOS BY THE AUTHOR

"WHERE ARE YOU?" my wife, Barbara, asked. I looked through the glass of the telephone booth, searching for a city-limits sign in the headlights of passing cars. What was the name of this place, anyway? Then I remembered.

"We're in Fancy Gap, Virginia."

There was a moment of troubled silence, and then Barb said, "Is that on the way from Road Atlanta to Wisconsin?" This was when we still lived in California, and her voice traveled across the continent.

"Well, not exactly," I admitted. "Chris and I decided to take the scenic route back to his place. I'll fly home later in the week. We're following the Blue Ridge Parkway up the East Coast until we hit the Martin guitar factory."

"Where's that?"

"In Nazareth, Pennsylvania. We'll only be a few days late. It's not really that far out of the way . . . "

"Yeah, right," I thought. "Only about a thousand miles."

Ah well, what else could we do? There we'd been, eating at Major McGill's Fish House near Road Atlanta after the Sports Car Club of America Runoffs, looking at a map and trying to figure out the best route home. Parked in front of the restaurant was my friend Chris Beebe's "new" shiny black 1956 Austin-Healey 100-4. He'd just traded a Westfield Eleven kit for the Big Healey, swapping straight across with a man named Mark Daniels, an old friend and my first SCCA driving instructor.

Our venerable Rand O'Malley off-brand map of the U.S., laid on the table in the fishy ruins of our smoked-trout dinner, showed a vast diagonal slash of topographic wrinkle called the Appalachian Mountains, and through it ran this convoluted trail of golden dots

called the Blue Ridge Parkway. If ever a line on a map looked like Sports Car Country Deluxe, this was it.

Every pilgrimage needs its Mecca (or Canterbury, for those pilgrims who favor SU fuel pumps), so I had diligently searched the northern terminus of the Parkway until I found one on the map: Nazareth, Pennsylvania.

Famous for its Andrettis, this little town was equally celebrated among guitar pickers as the home of C.F. Martin & Company, makers of fine American stringed instruments for 158 years. Chris and I both own Martin guitars and I'd learned through correspondence that Charles F. Martin IV is a sports-car buff and an R&T reader, so the factory seemed like a natural stop.

So on a Monday morning in October, we broke camp at Lake Lanier, Georgia; stuffed our earthly belongings into the many nooks, several crannies and low-slung trunk of the Healey; and struck out for the Blue Ridge, leaving Georgia and picking up the trail near Bryson City, North Carolina. Right at the southern edge of the Great Smoky Mountains National Park.

The Blue Ridge Parkway is a road designed purely to be scenic, built mostly in the Thirties under one of President Franklin D. Roosevelt's public works programs. Road engineers actually visited houses and cabins in the mountains and asked people where the best views were, then linked them with a road. The idea was to connect the Great Smoky Mountains National Park with the Shenandoah National Park in Northern Virginia. The Blue Ridge Parkway heads northeast for 469 miles and then becomes the Skyline Drive for another 105 miles through Shenandoah. Skyline Drive is supposed to have 70 scenic overlooks.

New country, new car. I'd never seen the Blue Ridge, nor had I driven a big 4-cylinder Healey. Experience with the marque was limited to Sprites and a few short drives in the 6-cylinder cars.

Big Healeys were not something I lusted after as a teenager, probably because I was born just a little too late to appreciate the virtues of the lighter, racier 100-4s. By the time my generation started noticing Big Healeys, they were being advertised as family sports cars. Nothing, of course, spooks a high school student quicker than the vision of two small children hanging around in the back seat of a cool sports car, which shouldn't have a back seat in the first place. "Family of Four Wins Le Mans" was a headline you never saw. The Healey 3000s seemed to be aimed at old guys over 30, the kind who spent perfectly good car money on house mortgages and lawn-care products.

Now, of course, I am an old guy and they seem fine. Chris's 100-4, however, appeared to be even finer; it was a car whose exact virtues I had somehow failed to notice.

We pulled over to trade places in North Carolina and I took my first drive. For about 40 feet. As I merged back onto the highway, the clutch arm under the car hit the road shoulder and bent, causing the clutch to slip. We stopped at a boat-launching ramp along the French Broad River, drove one side of the car up on a pile of rocks and adjusted the clutch. We made no jokes about the river's name, because that would be wrong.

Lesson One: Big Healeys are low. And their clutch arms are even lower than the bottom of the chassis. From

then on we winced and swerved at the sight of dead squirrels, no matter how flat, and entered driveways on a cautious diagonal.

Lesson Two was not long in coming: Big Healeys are also hot. Chris had insulated the driver's-side footwell with standoff shields of aluminum and asbestos before the trip, but the cockpit was still very warm, even in the cool autumn weather. The large exhaust header passes right under the floorpan and there's not much venting for the engine compartment, so the footwell on both sides becomes an English interpretation of the Dutch oven.

It's not tennis-shoe-melting hot, but I gathered we wouldn't be needing the heater unless the weather dropped to absolute zero and all molecular motion in the universe, as we know it, suddenly ceased.

Except for those slight inconveniences, the 100-4 immediately struck me as a marvelous road car, especially for 1956. The 2.7-liter (2660-cc) Austin inline-4, rated at 90 bhp, is remark-

ably smooth and pulls like the tractor from which it was derived, topping out at a 4800-rpm redline. It's so torquey at low rpm, however, that there's not much point in revving it above 4000, except to make Big Healey noise.

Unlike the earlier BN1 version of the 100-4, which had a 3-speed gearbox and a lockout plate over the stump-puller 1st-gear slot, the BN2 (3924 copies, made only one year) has a genuine 4-speed box mated to an overdrive unit. It's that overdrive, combined with the gutsy engine, that makes the car.

Hit 70 mph in 4th and you are turning a slightly busy 3600 rpm. Flick the overdrive toggle on the dash and it drops to a mere 2800. Eighty mph is only 3100 rpm and the car will (literally, for once) cruise there all day long, emanating a mellow cloud of low throaty exhaust music. On one empty stretch of road, we tried 100 mph. The tach said 3900 rpm. No sweat, no problem. Serenity.

"My God," I said to Chris. "This is a real car . . ."

Our last cross-country trip, I should mention, was made in a Citroën 2CV ("Au Revoir Vitesse," October 1988). Which is also a real car, but it's a real slow car. You can starve to death trying to reach the next cafe in North Dakota.

No such trouble with the Healey. We ate like kings. Entering the Parkway, however, we discovered there was no need for an 80-mph cruising speed. The smooth, winding two-lane road that climbed into the mountains was posted at 45 mph. While this sounds like a cruel restriction, it proved not to be during our first morning of driving. The scenery is so spectacular from the road that traveling much faster distracts from the constantly unfolding view; tire howl through the endless tight corners is somehow at odds with the misty, mysterious quiet of the vast forests and foggy valleys on either side of the ridge.

We stopped by the side of the road at

a lookout and watched the tendrils of an early evening ground-fog snake slowly through the hardwoods, moving upward and toward us. "I hate to say this," I told Chris, "but it looks just like the Central Highlands of Vietnam did at this time of day. Quiet fog and dense forest. Beautiful, but ghostly."

If there are any ghosts in the Smokies, however, they are phantoms from another conflict. These mountains were ancestral home to the Cherokees. Andrew Jackson had them forcibly removed to concentration camps and then to Oklahoma on the infamous Trail of Tears. The Cherokees, well organized and literate, tried everything to merge peacefully with American society, but it didn't work. They made the mistake of living on valuable real estate, though I doubt they called it that.

If you believe in historical resonance, it's very easy to imagine that a tinge of sadness remains in the mountains. To me it seems as if they're still waiting, like an empty house, for the owners to come home.

After a brief rain shower, we put the top down and motored along in a low growl through the soft evening air, stopping for the night at the small town of Maggie. There are almost no buildings or commercial establishments on the Parkway itself, but frequent exits allow you to drop down into nearby towns for fuel, food and lodging.

I suspect the Parkway is quite busy with tourist traffic in the summer, but in October the road is largely empty. You see another car or the random RV/truck about every 20 minutes. At times it takes on the existential aspect of an empty, private road continuously self-created in your path, just in time to be enjoyed, then disassembled right behind you. Sartre would have liked it, if he could drive.

It also rains often on the Parkway, and the Healey 100-4 is one of those legendary British roadsters, like the TR-3, whose weather gear was designed on the lobster trap/Roach Motel principle: Rainwater checks in, but it doesn't check out. It funnels through small openings and is then trapped in a larger chamber, known as the cockpit. Nothing seals; not the side curtains, windscreen gaskets or doors. As you drive, large droplets of water either dribble onto your knees or spit at you with pea-shooter velocity, wind-driven through cracks. It's like the OK Corral.

We stopped at many hardware stores, damming seams with rope caulk, tape-backed insulation foam, duct tape, etc., but nothing had much

effect. The viscosity of American water is too light for British cars.

Thick fog occasionally forced us off the Skyline Drive and onto major highways, where we made good time. Two days of driving got us through North Carolina and into Virginia, but low (actually, no) visibility drove us out of the hills and onto Highway 340 through the Shenandoah Valley. Names on the map of this green and peaceful area still ring of the Civil War. We drove through Front Royal and Winchester—which changed hands 71 times during the war—just over the ridge from Harpers Ferry and Manassas.

The relentless Stonewall Jackson double-timed his weary gray-clad troops up and down both sides of these mountains, crisscrossing the ridge over impossible roads and making life miserable for a long succession of less gifted and less audacious Union commanders. So did Lee. Lincoln was not amused. The whole Shenandoah Valley, in fact, is so rife with Civil War history that any stopping point could provide a reasonably detail-minded historian with a life's work. I shall not attempt here to further summarize this theater of the Civil War, however, as I would have to look things up. Check with Bruce Catton and Shelby Foote.

Our Skyline Drive tapered off and ended at the Rappahannock River, so we cruised across the tip of West Virginia into Pennsylvania on Interstate 81 and I-78 to Bethlehem, turning north on Highway 512 to 248. On a cold, rainy afternoon we rode into Nazareth, as the song says, feeling about half-past dead. A short search of the small town found us the Martin factory at the edge of town. It's a modern steel building (the older, more rustic factory is a short distance away). The new facility could just as easily be making plastic plumbing fixtures as hand-built guitars.

Until you get inside. There it smells like glue, hardwoods and sawdust, and everything is all right. The company historian and PR director, Mike Longworth, revived us with hot coffee, made us welcome and took us on a factory tour. C.F. Martin IV, he said, was off on a business trip and would be sorry he missed us and our Healey.

In the well-lighted factory workshop, the pace was quiet and unhurried, with craftsmen and women shaping guitar necks, bookmatching beautiful pieces of Indian rosewood, carefully gluing on herringbone bindings and mother-of-pearl fret markers, making the same legendary wooden boxes that resonated behind the voices of Gene Autry, Hank

Williams, Elvis, Neil Young and hundreds of others, beginning in 1833. I thought that if cars were made here, they would probably be Morgans. Healeys were a little too modern to come out of the Martin factory, while most parts of a Morgan could have been made in 1833, no problem.

We said our farewells, turned west and made tracks through the beautiful Appalachian ridges and valleys of Pennsylvania. Then it was overdrive time—Interstates through Ohio and Indiana, where high winds and driving cold rain kept our Healey busy collecting more water samples than Jacques Cousteau. It snowed briefly in Indiana, and we actually turned the heater on. It didn't work, of course, until we reconnected a couple of loose fan wires. Then it blew a wad of mouse bedding into our faces, and the vents showered us with rodent-scented confetti for the next few hours. Very festive. At the same stop, we added our first and only quart of oil to the engine.

At Chicago we finally drove out of the dreaded lake effect and into warm sunny weather. The Healey hummed up into Wisconsin, along the Rock River and into the driveway of Chris's farm on a Friday afternoon. I decided to stay the weekend before flying to California.

On Saturday, Chris's neighbors from across the creek stopped by for a visit. They said they were thinking of selling their old farm so they could find property with more pasture land for their horses. We got a guided tour of their old place on a lovely autumn afternoon, and I fell in love with it. I said I would ask Barb to fly out for a look. If she liked it, we'd buy it.

She did and we did. A few months later we picked up stakes, left California and moved in.

The old Healey still lives across the creek in Chris's garage. On warm, sunny mornings, he drives it to work with the top down and the windscreen tilted into its rakish, laid-down racing position. Sometimes he stops by for a visit on the way to work, and we sit out on the side porch drinking coffee, quietly looking at the car.

When Chris drives off, I always watch until the black Healey disappears into the woods and I can't see it any more. Then I stand perfectly still with my head tilted and listen until the sound is gone. If all car designers could make us wish to do only those two things, they would probably be as well remembered as Donald Healey.

I will always remember the Big Healey itself as the car that brought me home, with a slight detour on a good road. ◉

The potential-laden Super Seven frame.

The Monterey Weekend
TRACK TIME

Getting there is half the fun, or maybe a little less

BY PETER EGAN

PHOTOS BY RICHARD M. BARON & JOHN LAMM

96

■ Above, the author
races his freshly completed Super
Seven at Laguna Seca until . . . you'll have to read
the story. Below, a rare and esthetically questionable "Breadvan"
Ferrari 250 GTO. At left, a Talbot-Lago T26 spins its wheels in the pits.

197

■ At far right: Honoree Juan
Manuel Fangio and Froilan
Gonzalez autograph books; at near
right, Fangio's championship-
winning Mercedes-Benz W196;
and, above, the Lotus in tow.

J UST EIGHT TREMENDOUSLY long weeks before the Monterey Historic Automobile Races, Editor Tom Bryant called and asked if I could have my 1963 Lotus Super Seven restored and ready to run by then.

Tom said, "I talked to Steve Earle [who runs the whole thing], and he said there's still time to enter your car. I'd like a story from a participant's point of view. Can you have the car ready by the middle of August?"

"Oh, sure,
I said. "After all, I've
nearly finished sanding the frame . . ."

"Good. See you there."

I hung up and stared into middle distance, where impossible racing dreams dwell, right alongside patent applications for 200-mpg carburetors and the presidential aspirations of Democrats.

Eight weeks. Oh boy.

I looked around the garage.

Engine half apart, under the workbench. Spiders in the cylinder bores. A bare frame on sawhorses.

On the bright side, the frame was straightened and reinforced. Rear axle and front hubs were ready to bolt in. A new fuel cell, Tilton pedal assembly and radiator waited on the shelf, along with a new set of fenders. Wings, I mean.

On the dark side (here I made a list), the car needed the following: a new rollbar, padding for same, a floor and footwell, new aluminum skin, 1000 rivets, lower A-arms, paint, upholstery, brake lines, fuel lines, wiring, wheels, tires, seatbelts, instruments, dry sump tank and oil pan, oil hoses, oil cooler, windscreen, kill switch, Dzus fasteners, nose repair, grille, fuel pump, pressure regulator, instrument panel, brake pads, coolant hoses, catch tanks, header tank, valve job, pump switches, remote oil-filter housing, distributor and plug wires, sparkplugs, water pump belt, gaskets, anti-roll bar brackets, master cylinders, clutch slave and pedal assembly brackets. The driver needed a new helmet with a 1985 Snell sticker.

The car needed (I would later tabulate and discover) $4000 worth of parts. And the labor to bolt them together into a functional racing car.

To that end, my friend Chris Beebe agreed to help with the project. Together we would assemble the car at his shop, Foreign Car Specialists in Madison, and then tow it from Wisconsin to California. He, in turn, would charge me a small multiple of the national deficit, while still throwing in several hundred hours of free labor and giving up all vestiges of normal life for two months.

And so it began.

T hose eight weeks are something of a blur now: The Lost Summer.
Both of us generally worked until early afternoon on our other jobs, then labored until about 3:00 a.m. on the Lotus. We became night creatures, living on delivered pizzas ("You guys still open?") and tortilla chips and coffee from the all-night convenience store on the corner. I got my own giant insulated plastic VIP Coffee Club refillable mug. The late-night manager played loud rap music on a boom box—music by guys with Ice in their names, threatening to do all kinds of nasty things.

At 2:30 a.m., the other customers in a convenience store also look as if they have Ice in their names. Or somewhere.

One night I installed the Seven's floor and transmission tunnel, popping about 400 steel-mandrel rivets by hand. In the morning, my arm was swollen, black and purple from thumb to elbow. Torn muscles and blood clot, the doctor said. "Don't do any more rivets."

Yeah, sure. I got myself an Ace bandage and a rivet gun with longer levers.

Some nights we got carried away, worked too late, and the sun came up over the doughnut shop next door. On those mornings, we had a nutrition-packed breakfast from the Bear Claw and Bavarian Creme Turnover food groups.

We learned that a human male can get by on three to four hours of sleep a night and still have full recall of simple information vital to survival, like the exact date of his wedding anniversary.

We also rediscovered the basic law of race-car construction: Nothing Ever Fits.

For instance, three days before Zero Hour, we found that our oil pump, dry sump pan and the fittings that joined them were of three different diameters, thread styles and sealing systems. Another all-nighter. Some fabrication and welding skills required. Kit recommended for children ages 40 and older.

Friends helped. Tom Dula, Pat Donnelly and John Oakey pitched in and made the final difference that got us on the trailer in time. Our friend Denny Marklein painted the car in his body shop in one day, less than a week before Monterey. A flawless job. We almost left the car in unpainted aluminum because we couldn't bear to give it up for 12 hours.

Pat lent us his old race-car trailer, a rig with axles from a Renault R-16 and typical French 3-bolt wheels with skinny tires. We replaced the old bearings and tires. Could tiny bearings and spindles meant to hold up one end of a Renault carry a 600-lb. trailer and an 1100-lb. race car across the Great Plains and the Rockies? Or across town? I feared we'd find out.

O ne beautiful Monday morning in August the sun came up on our labors and I said, "We have to leave now. We have exactly 72 hours to be at tech in Monterey, which is 2400 miles from here."

Undriven, untested and still missing its windshield and numbers, the car was pushed onto the Renault/trailer. We lashed it down, loaded tools, spares and fluids into my Chevy van and towed 22 miles south to my house, where we switched towing duties to R&T's longterm Ford Explorer. The only reliable wheeled object, you will notice, in this whole picture.

On the way to the farm, the Chevy blew a heater hose, but I kept driving. "No time to stop, " I said. We sailed up the driveway in a cloud of steam, with the engine knocking. This is what my friend Larry Crane calls "Pressing on non-irregardlessly."

The trailer, of course, was tongue-heavy, pushing the van all over the road. We spent two hours moving the rusted wheel chocks rearward on the trailer, took short naps to fend off total incoherence, fired up the Explorer and sped off in the general direction of Iowa at 3:00 in the afternoon.

Drive drive drive drive drive. Sunsets, sunrises. Coffee, burgers, gas. Short stops for sleep in Cedar Rapids and Wendover, Utah. Sunrise over the Salt Flats. More driving. Rush hour in Sacramento, traffic spreading into the surrounding hills like venom from a snake bite. We crawled into Carmel at night, found the motel. Chris fell asleep in his chair, cutting out four sets of Con-Tact-paper numbers. Ninety-six, my assigned number.

An early morning drive over Laurales Grade Road in light fog got us to the track. A man at the gate directed us straight to our numbered slot in the paddock. We unloaded the car, fired it up and I drove to tech.

First drive! It ran and shifted. Even the steering seemed to work. The brake pedal caused the car to slow down. Neat stuff.

The tech inspector looked at my engine and asked, "Is that a dry-sump oil system?"

"Yes."

"A 1963 Lotus Seven didn't come from the factory with a dry-sump system, did it?"

"No," I said. "But I was here two years ago and there were several Sevens with dry sumps. I thought I'd be foolish not to have one."

"They're not legal anymore." He called another tech inspector over. "Hey, can this guy race a 1963 Seven with a dry-sump system?"

"Nope."

There was a long silence. "You can't race it like this," he said.

Oh boy. Eight hundred dollars' worth of components and two extra days and nights of work, just to get disqualified in tech. "We don't have the parts to change it back," I explained. "We just got here, all the way from Wisconsin."

More silence.

The first tech inspector called Race Central on his radio and explained the situation. There was a slight delay, and then the disembodied voice of Authority said, "Let him run this time, but tell him never to come back here with a dry-sump system again."

We were in.

On Thursday morning the Jim Russell folks ran a drivers' school for those who, like me, had never driven the revised Laguna Seca circuit with its new infield. So I went through the school and discovered a few things about the Lotus.

First, it had terrible bump steer and was very twitchy, darting everywhere under braking. Our quick-and-dirty 11th-hour suspension setup left a little to be desired. Once in corners, however, the Lotus was well-balanced, neutral and easy to drive. Fun, even. The Vintage Dunlop tires ($150 each, via MasterCard) were sticky and forgiving.

The engine, despite being a nearly stock 1500 Cosworth-modified Ford, had reasonable torque and ran beautifully. Up to 6400 rpm of its 7000-rpm redline. There, point bounce took over.

■ 1961 or 1991? Sir Jack Brabham again in the Cooper-Climax T-54; below, a Lotus 26R.

"Have to fix that this winter," I said. "No time now."

Also, there was no mechanical stop on the clutch pedal (no time, etc.), so we adjusted it to release near the floor, to prevent over-center damage to the pressure plate arms.

With this weird combination of problems, we still managed to qualify 10th out of 24 cars, mostly as a credit to the Seven's good power-to-weight ratio and low center of gravity. I found myself, in Group 3B, surrounded by Morgans, Alfas, Renault-Alpine A110s, Elva Couriers, Porsche 904s and a very nice Ginetta, whose tail I clung to while trying to learn the new course.

At the front of our group was a small pack of the fastest Lotus 26R Elans I've ever seen. They were in a race of their own, passing the rest of us like Can-Am cars in a bicycle race. My friend Gil Nickel was driving a yellow one, and he waved as he passed me. By the time I waved back, I think he was too far away to notice.

"People who don't think vintage racers drive fast ought to get out here with

■ From top: A Porsche 910; a Talbot-Lago T26C overtakes an ERA; and Steve Earle's Alfa GTZ-2 leads two Alpine-Renault A110s.

this bunch," I mumbled into my Nomex balaclava, straining to keep up with Robert Forbes's Ginetta.

There was only one other Lotus Seven, a beautifully prepared burgundy car driven by Ron Bennet. The car was faster than a bullet and well-driven, so I didn't see any classic wheel-to-wheel battles looming on the horizon. He was several rows ahead of me on the grid.

Our practice and qualifying sessions quickly reminded me that one of the truly appealing aspects of vintage racing is the sight of the other cars on the track. Visually, few things are lovelier than a close-up view of a couple of Morgans, a few Elans and a 904 drifting through a corner on a sunlit track—even if you're still behind them. It's like having a front-row seat (or, in my case, a middle-row seat) at a fast-moving concours. With all the right smells and sounds. This is how sports cars are meant to be seen. I recommend it.

When Friday's practice and qualifying were over, we had Saturday to tinker with the car and wander around the paddock. This year there was, for a

the crowd atmosphere was one of hushed reverence with a great deal of electric current running through it.

Chris and I got Fangio's autograph and shook his hand. So did about a thousand other people, but the great man with the clear, piercing eyes managed it all with smiling patience and dignity. He turned 80 this year and still doesn't wear glasses, as some of us I can think of now do. He looks as if he can see through an engine block at 300 yards.

In any event, there's nothing quite like shaking hands with the man you believe is the greatest racing driver who ever lived.

Later, Fangio got out on the track with the open-wheeled Mercedes 196 and did several respectfully fast laps. As he came downhill out of the corkscrew, a camera car that was pacing him spun out. We stopped our fiddling with the Lotus and stood on a trailer to watch the 196 and listen. What a sound. One of the last real Grand Prix cars. Don't ask me what "real" means. You have to see these cars on the track, look closely at the craftsmanship when they're in the

a green and yellow Lotus 26R when my clutch pedal clanged uselessly to the floor as I exited the downhill Corkscrew. Blown clutch hydraulics. Oh, well, put her in gear and drive clutchless. Nope. No power to the gearbox.

I coasted into the pits, and we discovered the clutch slave cylinder piston had popped out, jamming the pushrod and disengaging the clutch. By the time we pried it loose, the checkered flag came out. A 26R driven by Dirk Layer won, followed by Gil Nickel's version of same.

As we pushed the car toward the paddock, the flagman kindly reached over the wall and waved the checker for us.

It was a disappointing finish to the weekend, but at least we'd made it into the race. And we'd still had three good sessions. Drivers' school, practice and qualifying. Nearly two hours of track time, on a car that had never turned a wheel before it arrived at Laguna. We had fun and the car was still in one piece. It could have been worse.

After the last race on Sunday, there was champagne for all and a fine speech

■ At speed is Robert Rubin's 1929 Miller 91. On static display is a Lancia-Ferrari D-50, the sort that Juan Manuel Fangio drove to his fourth World Championship in 1956.

change, no featured marque. Instead, it was a year to honor Juan Manuel Fangio and to exhibit many of the cars he drove to victory.

With Fangio was his old compatriot Froilan Gonzalez. They strolled through the paddock together, looking at the remarkable display of Fangio's winning cars: a Maserati 250F, two Mercedes-Benz 196s, a Lancia-Ferrari D50 and an Alfa Romeo 159 Alfetta. Behind the cars were huge photographs from the era, by Jesse Alexander and Louis Klemantaski. The combination of photos and cars with these legendary drivers standing nearby was stunning;

pits, hear the sound of the engines, watch someone like Fangio drive them. It all adds up.

Saturday night a bunch of us ate Japanese food and drank sake. Parts of the evening were very funny, as I recall.

Luckily, my race wasn't until Sunday afternoon. Chris and I arrived well before noon and managed to overheat the engine while trying to reset the timing, just before my race. We had to shut the car off and let it cool, pushing it to the grid.

The 10-lap race was wonderful, for about 3 laps. I was working hard to pass

by Fangio. With the red sun dipping into the Pacific, we loaded up our car and tools and headed for home.

It was a leisurely return trip. We ate full meals, slept eight or 10 hours a night and made it back in four days. Nine miles from home, a leaf spring broke on the trailer, and we pulled into my driveway at night with the trailer clanking and listing heavily to the right. No harm done. Home safe.

A Lost Summer?

Not really. The Lotus, which used to live in rafters and shelves and moldy cardboard boxes, is now consolidated into a compact working unit of British Racing Green. Furthermore, it actually raced at Monterey. The French trailer wheels didn't fall off, and we shook hands with Fangio. Our wives, Barb and Dana, have begun speaking to us again. And we have the enduring memory of those vivid, slightly surreal sunrises over the doughnut shop on hazy summer mornings. That seems enough to ask of one short season.

THE GREAT GP TREK

For two young Americans in Paris in 1971, the only thing standing between them and their first Grand Prix was a 1000-mile bicycle ride to Barcelona

BY PETER EGAN
PHOTOS BY THE AUTHOR

O KAY, WHERE ARE we going on this big bicycle trip?" my friend Bill Steckel asked me, exhaling the usual cloud of Gauloises smoke and signaling the waiter for two more *ballons de vin rouge.*

"Barcelona," I said, tossing a French car magazine on the table. "The race schedule says the Spanish Grand Prix is coming up in April, at Montjuich. By that time, the weather should be pretty good in the Pyrenees. We could get a couple of bikes and ride them down there. Take maybe two weeks. I figure it's about a thousand miles."

We were sitting indoors at a bistro in Paris, a place called La Chope, on Place de la Contrescarpe. The sidewalk tables were glassed in for winter, and just outside we could see *les clochards,* the famous winos of Paris, sleeping over the warm sewer grates with steam escaping around them.

At the bar a bunch of male Sorbonne students were arguing with muted savagery over some political problem, all of them dressed in tailored wool bell-bottoms, expensive shoes, shag hairdos and short leather jackets. The Sorbonne, in 1971, produced the best-dressed revolutionaries in the world.

Bill Steckel and I were both bumming around Paris for the winter. I had recently gotten out of the Army, staying home just long enough to convert my $1100 of accumulated combat pay into a small wad of traveler's checks and an airline ticket to Europe. Bill was a business school graduate who'd managed a trucking company for two years and then decided to take his savings and travel.

We'd met at the Hotel Victoria, near the Panthéon. The Victoria was a Left Bank hangout for the $5-a-Day crowd, at a time when that was still a realistic figure. I lived nearby at the Modern Hotel, which had been modern about the time of Napoleon's visit to Marengo. My room cost seven francs a day (about $2) and was shaped like a small slice of pie, defying rectangular furniture. The window looked out on an air shaft with a patch of sky at the top.

Bill was a big shambling guy who chain-smoked Gauloises and wore sport coats and bow ties at a time when the rest of the world's Youth Culture affected a kind of Basque shepherd/Marrakesh rug-dealer look. His Ivy League orthodoxy made him the only true eccentric in the hotel, and everyone was charmed by his easy good nature.

It was a great time to be in Paris. There were a lot of shiftless, good-time Americans around, keeping up the ancient bohemian tradition. During the day you could walk through the Louvre or the Jeu de Paume galleries unmolested by tourist crowds, or walk down the Seine to Shakespeare & Company, or look at the book stalls and drink coffee in the cafes.

At night a bunch of us would get on the Metro, ride to a distant restaurant, such as Julien at the arches of St. Denis, and then drink our way home, bar-hopping from one corner bistro to another.

During one of these sybaritic evening hikes, I discovered that Bill had ridden a bicycle across the United States, from Seattle to New York, when he was 18. I'd never taken a long bicycle trip, but the idea was appealing. So we decided to bicycle somewhere in Europe at the first sign of spring. The Spanish Grand Prix beckoned.

Buying a bicycle in Paris was not easy. We'd look up the address of a bicycle shop in the phone book, go there and find a narrow little store with a wall full of sprockets, alloy tubes, wire wheels and racing hubs. The proprietor would approach us with tape measures—wanting to check our inseams and sleeve lengths so he could *build* us a bike. For about $600. Thank you, no,

■ Above: Brake pads gone, 15 pounds thinner, Steckel (left) and Egan arrive at the Spanish border. Right and below: Teammates Ickx and Andretti in the Ferrari paddock. Lower right: Rainy practice at Montjuich Park. Bottom: Graham Hill serves lunch.

monsieur, we have made a mistake.

Finally we found a big Peugeot factory showroom on the Champs Élysées, Bicycle City. We bought a couple of 8-speed touring models with lights, fenders and luggage racks, for about $80 each. Perfect. We rode them back to the hotel in rush hour, around the Arc de Triomphe, and lived. A good omen.

Early on a Monday morning in April, we said goodbye to Paris and hit the cobblestones, each of us with about 40 lb. of worldly belongings on our luggage racks. Army surplus sleeping bags and shelter halves—courtesy of the flea market—were strapped to our handlebars. My old Nikon F was slung over the bedroll like a headlight. Our first night's objective was the youth hostel in Chartres, 60 miles from Paris.

On this ride, Bill stayed in lower gears, pedaling furiously, while I stretched it out in top gear. Neither of us had ridden a bicycle in five years, so Bill advised me to shift down and save my legs. "Too much work," I said.

We stopped for a look at Versailles and then approached Chartres, with the famous Cathedral towering in the distance like a ship on a sea of wheat fields. We reached the youth hostel at sunset. We had a typical youth hostel dinner of collectivist gruel, wrote a few postcards and went to bed in the dormitory.

At midnight I awoke with my legs in spasm. Hundreds of small hand grenades were going off in my muscles. I couldn't straighten my legs, so I rolled out of my bunk, slammed the floor and crawled down the hallway to the shower room, hoping to run hot water on my legs. When I got there, I remembered two important things:

1. There is no hot water in French youth hostels.

2. Showers are operated by an overhead pull-chain. Unreachable, if you are a lowly quadruped.

So I crawled back to bed. Not having a gun to kill myself, I was still there in the morning.

■ ■ ■ ■

The French people in small towns are warm and generous. Unless they drive cars. There is an invisible force field at the transom of every French car that turns perfectly nice people into murderous psychopaths.

■ ■ ■ ■

We lost a day, as Bill taught me to walk again, then hit the road toward the Loire River valley. I shifted down and stayed there.

The second night, we pitched our tent against a tennis court fence in a park along the river, at Meung sur Loire. We woke up damp and sore, feeling generally like reprocessed dog meat (meung?), and vowed to stay in cheap hotels and youth hostels rather than camp again.

On the road, there were more lessons to be learned, the first being geographical.

France, it seems, is a tilted country, with the high end in the south. This large, tilted block of real estate is called the *massif central* and runs uphill all the way to the Pyrenees on the Spanish bor-

der. It gets steeper as you go south. And, in the spring, higher and colder, terminating in a border region that looks positively Canadian, with pine trees and snow. Nelson Eddy/Jeanette MacDonald country.

Another lesson was cultural.

The French people in small towns are warm and generous. Unless they drive cars. There is an invisible force field at the transom of every French car that turns perfectly nice people into murderous psychopaths. Presented with a natural victim like a bicycle on the road, French drivers like to honk, flash lights, swerve and make extravagant Gallic gestures of lip and hand. At least they did then. So we took the most remote, untraveled roads we could find, south through Bourges, Montluçon, Clermont-Ferrand (not so remote) and St. Flour.

We also learned not to eat "lunch" on the road. Even at truck stops, the French *déjeuner* tends to be a five-course affair with wine. It leaves you feeling happy, numb, sleepy, fat and nearly incapable of finding the men's room, let alone balancing a bicycle. Especially on a road filled with truck drivers who've just had the same lunch. We cut back to light, high-energy lunches of Suchard chocolate and the famous *beaujolais American*: Coca-Cola. Carbohydrate loading became a much-anticipated evening ceremony. Also meat, wine, cheese, dessert and cognac loading. We loaded it all and still lost weight.

Speaking of health matters, I had to hand it to Bill; as we climbed higher into the mountains, breathing thinner and colder air, he never stopped smoking his Gauloises. I'd look over as we rounded an uphill low-gear hairpin and see him trailing small puffs of smoke, just like a train. He would occasionally pedal up beside me and suggest we stop

■ Left: Sundown on the first 60 miles; Bill and bikes at Chartres. Right: We left our bikes unlocked everywhere, but no one would steal them.

for a cigarette break.

As the weather grew colder and the roads steeper, we began leaving our bicycles unlocked in front of hotels, in hopes someone would steal them. But every morning when we came out, they'd still be there. "What's wrong with these people?" Bill would ask. "Are they blind?"

Our worst day of the trip was a climb between St. Chéle and Millau, a tortuous uphill flog into a stiff mountain gale and sleet storm. We were in low gear, wearing our rain ponchos, when a truck sped by. The wind blast blew our ponchos up over our heads and stopped us dead in our tracks. Blinded, feet stuck in pedal traps, we fell over sideways, like toppled equestrian statues. Bill's cigarette was ruined in this crash, which called for a smoke break in the shelter of a nearby barn.

After about 18 miles of progress for an afternoon's effort, we stopped for the night at the windswept little village of Millau. At the hotel desk, we trotted out our favorite question: "Are there many hills between here and ------- [fill in the name of the next town]?" In this case, Lodève.

"Mais, oui!" the concierge assured us, eyes growing large and frightened, like those of a superstitious innkeeper discussing werewolves in a Peter Lorre movie. She assured us it was impossible to get to the next town by bicycle and made uphill gestures of the arm, which I judged to be about 85 degrees in grade. Take the train, she told us.

There's one tomorrow.

When we came out of the hotel in the morning, it was still raining and our bikes remained unstolen, despite our best efforts to park them at an appealing angle. Should we give up and take the train?

No. We decided to pedal one more day, then take the train if things didn't improve.

A few miles out of town we crested a beautiful gap in the mountains, a pass with cascading waterfalls next to the road. Ahead of us, the road ran downhill for as far as we could see along the rim

of a massive, beautiful valley full of vineyards, spring flowers and bright sunshine. Yet behind us it was still foggy and raining. A view into the South of France, like the gates of Heaven.

We coasted 80 miles downhill that day, through Lodève and Pézenas into ever warmer sunlight, never pedaling a stroke. After nine days of low-gear workout, we were gliding down like hawks on a thermal. All the way to the Mediterranean coast. When we got to Béziers, our brake pads were almost gone.

By following the coast, we skirted the

worst of the Pyrenees and the ride down to Barcelona was easy. Both of us had lost 15 pounds on the trip, and we were in great shape, able to pedal in high gear all day. After nighttime stops at Perpignan, Figueras and Malgrat de Mar, we breezed into Barcelona on a sunny Wednesday afternoon, our 13th day of riding, We found a room at a cheap hotel just off the Ramblas, a wide, tree-lined pedestrian boulevard. Many women lived in the hotel, but they kept strange hours and seemed to come out only at night. Perhaps they were vampires.

Restaurants nearby served a strange but hearty combination of roast chicken, chicken soup and Spain's interpretation of cheap champagne. Out on the Ramblas, elderly chaperones strolled with teenage girls while Spanish boys looked on, the braver ones politely introducing themselves to the chaperones and joining in the stroll. Just like traditional American courtship as we know it, only without cars, privacy, telephones, pizza, movies, loud music and sex.

On Thursday, Bill—who was not a big racing fan—went to Gaudi's famous architectural wonder, the Church of the Holy Family, while I took a bus to watch Formula 1 practice at the Montjuich circuit, which is in a beautiful park up in the hills of Barcelona. The buses were packed, but when I got to the track, no one else got off.

They were all going to bullfights or the football championship-cup games. There was hardly anyone at the track. Just the teams and drivers, with a few random spectators wandering around, like an SCCA regional. The uniformed *Guardia Civil* allowed me to stroll right into the paddock with no ticket and no press pass.

Unreal.

But there I was, dodging a rental car as Jackie Stewart and his wife, Helen, drove up and parked next to the Tyrrell-Ford van. Stewart was in his rock star phase, at the peak of his powers, dressed in the famous Jackie Stewart hat and *Superfly* sunglasses, with plenty of hair to get the goat of crusty old FIA types. François Cevert showed up and shook hands with Stewart. Nearby, a young Jacky Ickx watched his Ferrari being unloaded. Mario Andretti suddenly appeared, standing right next to me, also watching the cars come down the ramps. Andretti had just won his first F1 race, at South Africa.

This was all impossible. I was sure someone would throw me out at any moment, but no one did. Not all day. Nor during the next two days of qualifying.

I wandered over to the Brabham trailer and looked at the new "lobster-claw" car with its odd front ducting. A small crowd stood around the team motorhome, waiting for Graham Hill to come out. Eventually he did, emerging with a towel over one arm and a tray of small sandwiches held high. He passed sandwiches out to the mechanics without saying a word, holding his chin up like Jeeves the butler, then disappeared into the motorhome. Everyone laughed and applauded.

Just across the way, Rob Walker and John Surtees conferred with the Brooke Bond Oxo Surtees team's other driver, Rolf Stommelen. I stood next to them while they talked, like Woody Allen's *Zelig*. Ingratiating myself, chameleon-like into every historic tableau. Surtees, World Champion in cars and seven times on bikes. Should I say hello? No. It might break the spell and they'd throw me out. I decided that God had made me invisible, just this one day, as a special favor in compensation for my inability to play the trombone. Why wreck it?

It rained briefly late in the afternoon on Saturday, and everyone covered the cars with tarps. I walked past a car with its door open, and there was Pedro Rodriguez, watching the weather and waiting. He looked up and smiled. "Going out again?" I asked.

He shrugged. "I hope later the track will dry."

"Good luck," I said. He nodded graciously and autographed my race program, next to Andretti's signature. The paper was wet and the pen would barely write.

Rodriguez, who was going very fast in his Yardley BRM, would live to race in only three more Grands Prix. He was killed in a Group 7 race at the Norisring in Germany three months later.

Looking at the program now is a little sad. Besides Rodriguez, we would eventually lose Ronnie Peterson, François Cevert, Jo Siffert, Rolf Stommelen and Graham Hill from the starting roster.

There is another revelation in the program too: I see in the list of drivers for the Formula 3 support race the names of James Hunt and Alan Jones. I hope I didn't elbow these guys out of the way to get a better look at Andretti's Ferrari, or ask one of them to hold my coat while I took a picture of Stewart.

Race day was terrific. Sunday, April 18, 1971.

After missing the start because there was only one ticket booth at the park gate, Bill and I found a good spot for

■ Stewart, at the end of another brilliant drive, about to take the checker. Left: Future magic: Chapman and Fittipaldi. Below: Hill's "lobster-claw" Brabham.

watching the race. It was on a wall, looking down on the uphill S-bend just past the Calle Méjico. Cars would come screaming in from the short bottom straight, then drift under hard power up through the curves, with much wheelspin and tail-wagging.

Ickx's Ferrari led the first few laps, but Stewart's blue Tyrrell got by him and gradually pulled out a few seconds' lead. Late in the race, Ickx pushed harder and harder to catch Stewart, who drove one of the great races of his life to hold the lead and win. A first for the Tyrrell.

Looking down at the cars in the S-bend, I watched Stewart's tires just skim the Armco, first on one side and then the other, lap after lap, as he put the power down, howling and slithering up the hill. Each time, I would say to myself, "He can't get away with that again," but he always did. For nearly two hours. It was one of the most masterful displays of car control, bravery and concentration I've ever seen. Ickx, too, was tremendously fast, but I think his 12-cylinder Ferrari had a little more power than Stewart's V-8. Stewart was simply inspired that day. And on many other days, as it turned out.

Chris Amon finished 3rd, after a great drive in his Matra-Simca, and Pedro Rodriguez was 4th. A fine day of racing—classic battles in a beautiful setting—and a wonderful first GP. I didn't know it then, but I would never again have so much freedom to wander around a GP circuit or its paddock. Not even years later as a working journalist, with full press credentials. Spain was a dreamlike aberration.

After that weekend, Bill and I sold our bikes to a Barcelona bicycle shop for $40 and took a night train to Montreux, Switzerland, on our way back to Paris. We were supposed to meet some friends there, but they didn't show. So

Bill and I checked into the youth hostel and climbed a nearby mountain.

We struggled all day to climb the peak and got to the top only to discover there was a parking lot full of tour buses on the summit and a paved highway down the backside. We decided not to plant the American flag, or notify *National Geographic*.

Another train took us back to Paris, where summer tourist crowds had already descended on the city, turning waiters and bartenders surly for the season. Homing instinct and poverty hit us simultaneously, so we packed up and flew back to the U.S. In New York, Bill lent me $40 so I could buy a Greyhound ticket to Madison, Wisconsin.

Two days later, after nearly three years of living out of rucksacks, backpacks, panniers and duffel bags, I was home. Barb and I were married later that summer, and I went back to school.

When I think back on this trip, the most remarkable part is that I almost didn't take it. With $1100 of Army separation pay burning a hole in my pocket, I came very close to using the money as a down payment on a car—a brand-new MGB.

Now, 21 years later, you can still buy an early Seventies' MGB; we have a nice one in our garage, and I see others for sale nearly every day in the paper. Cars we can always find, and they are easy to restore. Good times with young friends in Paris, however, and bicycle rides to the 1971 Spanish Grand Prix have become much harder to arrange. ◉

Continued from page 113

that contains the shops and garages of Junior Johnson & Associates. Just up the hill from the shops, past the pens of hunting dogs, is the home of Junior and Flossie.

We get a tour of the various garage buildings—body shop, engine room, chassis shop—and then go up to the house with Tom. Flossie comes to the door and invites us in. She's the sort of person you immediately take a liking to—warm and friendly, but with a quiet personal dignity. She and Tom talk about the race suspension, and it's plain that she's very upset. A moment later, Junior walks in, wearing Oshkosh bib overalls and a white T-shirt. He's been out picking up a tractor, to help with the haying.

Junior's presence fills the room. He's a big guy—just under 6 ft. tall, 235 lb., according to the team press kit. The wrinkles around his eyes reflect a gentle, relaxed good humor, though the eyes themselves have a penetrating quality. It's the face of a person who focuses clearly on everything he looks at and possibly shoots a couple of X-rays before looking casually away.

Today the eyes look tired. "Bill France's office just called," he tells Flossie. "They told me not to come to the track on Sunday. Not even to watch the race."

Flossie is literally wringing her hands and her eyes are tearing up. "How can they do this to us? It doesn't make sense . . ."

"They said they want the press to pay attention to the race, instead of talking to me. They said to stay home." There's a catch in Junior's voice, a thickening of the tongue.

A long silence as Junior and Flossie look at each other. "I just can't believe this is happening," she says, "after all these years of racing . . ."

I want to beam myself out of the room. After a lifetime of reading and hearing about Junior Johnson, here I am standing in his living room at the moment NASCAR has apparently decided it should try to take him down a notch. I can only hope that it isn't as bad as it seems and that it's Chinatown again, and I don't understand what's happening. Still, it doesn't seem good.

Junior says goodbye and goes back to his haying, and Flossie invites us to come downstairs and see Junior's den and trophy room. It's a big basement party room with shelves and walls covered with silver cups, plaques and trophies. The dates go back 38 years. Flossie makes some of her famous biscuits and steak, with Bud-weisers all around.

As we eat, I look at the many plaques on the wall. The one nearest me is a Presidential Pardon: Doc. #4265W, violated section 5601, title 26. Nov. 27, 1956, 2 years, $5000. Pardon 1985, for Robert Glen "Junior" Johnson. Signed, Ronald Reagan.

It's a glimpse backward into another American age. Corporate grooming and professional drivers' schools will never produce another Junior Johnson. He comes from a time when greatness was a wild card.

On the way back to Charlotte on Highway 150, we pass a large weathered barn by the road. Tom points it out and says it was used as a set in "that terrible movie." He's talking about *Days of Thunder.*

He says people here opened their garages and homes to the movie crews, hoping something real would come out of it. "Now," he says, "they feel tainted by it."

Hollywood strikes again.

Racing. That's what we came here for.

In Saturday's 300-miler, Dale Earnhardt fights a tremendous battle for the lead with none other than Dick Trickle of Wisconsin Rapids. Trickle was a local racing hero when Pat and I were in high school, so we cheer him on.

Earnhardt wins, however. He's a tough guy to beat and a remarkable driver to watch on the track. Leading, he gives ground grudgingly; while being passed, he intimidates with small feints and movements of the car; and when following others, he fills their mirrors with lots of black paint from his No. 3 Mr. Goodwrench Chevy. A hard driver, the Man in Black.

And for the main event, the Coca-Cola 600 on Sunday, he's back. This time with only a little less success.

The race is preceded by one of track owner O. Bruton Smith's spectacular pre-race shows, this one a slightly confusing tank and ship battle inspired by the Gulf War. The race, too, will be slightly confusing. Not a vintage year for the 600.

Early in the race, about half the cars dive into the pits for fuel and tires under the green, having given up waiting for a yellow flag. Just as they are leaving the pits, the yellow comes out. This puts them a lap down on the other half of the field. It's a lap they never make up.

In most racing, this would have been the end of all drama. In NASCAR, however, yellow caution flags tend to bunch up the field when things get dull, so you can have a dozen or more restarts in a long race, all of them nearly as loud and exciting as the start itself. No matter what's happening in the race, it's a good show, like a prizefight with self-contained rematches.

In this race, one car stands out in every rematch. It's Davey Allison's black-red-yellow Havoline/Thunderbird. Every time he comes out of the pits in 4th or 5th place, he's back in the lead 4 laps later, passing other cars with impunity. Even Earnhardt can't hold onto him. Allison wins the Coca-Cola 600.

On the victory podium, the announcer asks Earnhardt about the race and he says, "Well, you can't outrun an illegal car like Davey's."

Seems Ford wasn't winning many races this season, so NASCAR allowed them to raise the rear decklid slightly for better downforce and handling. Adjustments like this are made all the time, of course, to keep the show in balance, to keep the people and the teams coming back for more. Allison counters that his car was dead-legal under the rules and that Earnhardt had not considered his own car "illegal" when it easily beat the Fords in previous races.

All good fuel for the great Ford-Chevy question, which I hope will be with us forever.

After the race, Allison leaves the victory podium and walks up through the grandstands to the press box for his post-race interview. He makes his way through the crowd, shaking hands and drinking a Coke, with his driver's suit peeled down to the waist. No bronzed god here, showing off. Thin, pale, dripping with sweat, tired and happy, he could be any one of the fans who shake his hand. An ordinary-looking person who happens to drive the wheels off a stock car.

In the press conference, he's quick-witted, candid and thoughtful. He says he awakened this morning not feeling well, sick to his stomach. When he got to the track, they were pulling his engine out because the valve guides were disintegrating into small pieces. Then he felt even sicker.

Someone asks how it felt to have a car that was running so well he was able to take control of the race at will.

He smiles wearily, seeming to contemplate the question with both irony and pleasure. Finally he says, "These times don't come very often, so we savor 'em while they're here."

Nicely put, with implications that go beyond the race itself. When you drive home from Charlotte through the Blue Ridge Mountains, the words resonate for a long time. ◉

In Search of the Blues

There are easier ways to get to Beale Street than in a 1958 Isetta, but very few that offer so much opportunity for bad luck and trouble

BY PETER EGAN

I F THIS SAME cold autumn wind had been blowing through Chicago, they would have called it The Hawk—that famous lake-driven breeze that sweeps through the skyscraper canyons and sends a damp chill straight to the bone. We were not in Chicago, however.

We were in Rockford, Illinois, where, as far as I know, the wind has no name at all. Nevertheless, it was plenty cold and we were sitting in an aubergine-maroon 1958 BMW Isetta bubble car, watching leaves swirl around the street while we waited at a stoplight. The light, in fact, had just turned green, and there was a lot of honking behind us.

After a few moments, I looked over at my friend Chris Beebe, who was driving, and said, "Why don't we go?"

He looked back at me and said, "I think the throttle cable has slipped. I'm pushing on the gas pedal, but nothing's happening."

"Are you sure?"

"Of course I'm sure."

"You got the right pedal?"

"Yes. I have the right pedal."

I slid my side window open and listened to the air-compressorlike exhaust note of the single-cylinder BMW motorcycle engine behind us and said, "Maybe it just loaded up the sparkplug at the stoplight. I think it's responding now . . ."

Chris slowly put his head down on the steering wheel and started to laugh. Then I started to laugh too. Pretty soon we were speechless, wheezing with the kind of laughter you normally hear among people who have misplaced their Thorazine prescriptions. We were pretty tired.

"Only in an Isetta," Chris said at last, "could you have a serious discussion as to whether or not the throttle cable is connected."

The honking behind us persisted, so we opened up the front and only door of the car, stepped out into the cold street and pushed the car into a nearby driveway. We lifted one side up onto a curb so we could look under the car. Chris fiddled with the throttle cable, which had indeed slipped through its rubber retainer sleeve, and I looked at my watch with a flashlight.

"It's over," I said. "It's now officially too late to make it to Road Atlanta for

PHOTOS BY THE AUTHOR

the Runoffs. Even if we drive all night and nothing else goes wrong and we average 45 mph, we'll miss the last race on Sunday. Let's find a motel, get some sleep and figure out a new destination in the morning."

Chris nodded, caving in at last.

After five weeks of the usual relentless frame-up restoration—eating Doritos, losing sleep and waiting for the UPS truck to deliver rare parts from diverse corners of the universe—we'd lost our race with the clock by mere hours. For the first time in almost 20 years, we would not be going to the SCCA National Runoffs at Road Atlanta.

We found a nearby motel, checked in, unloaded our nominal luggage and collapsed on our beds. I can't remember if I took my shoes off or not. Meanwhile, outside our door, the Isetta sat between a Taurus wagon and a Chevy Caprice as if posing for one of those "What's wrong with this picture?" contests.

A fair question. What *was* wrong with this picture?

Why an Isetta?

Well, because Chris used to have one in college and thought it would be fun to have another one. He found this one, disassembled, in northern Illinois, bought it for $1000 and dragged it home. Chris and his wife, Dana, carried the Isetta into their basement, placing it next to the washer and dryer, to whom it appeared related. The car sat there for

two years, until this fall, when I foolishly said, "We should rebuild that thing and drive it to the Runoffs at Road Atlanta next month."

So much for why. A more critical question in some readers' minds may be *what?* (Or, more accurately, "What the . . .?") As we would discover on this trip, some folks knew all about Isettas and had even owned them, while others had never heard the name and were virtually dumbfounded.

BMW Isettas were produced in Germany from 1955 to 1959, but the car was designed in the early Fifties not by BMW, but by Renzo Rivolta, a refrigerator manufacturer from Milan, which probably explains the front door and the heating system—more on this later. Diversifying from his line of Iso appliances, Rivolta decided to make scooters and small cars. The little egg-shape Isetta, originally powered by a 2-stroke engine, was essentially a kind of glorified Lambretta or Vespa, a scooter with weather protection.

Meanwhile, on the colder, darker side of the Alps, Germany was still digging its economy out of the wartime ruins, BMW was having a hard time selling its upscale 503s and 507s, and the Suez crisis had made fuel economy seem like a fine idea.

So as a stopgap measure to keep the assembly lines running, BMW bought production rights to the Isetta, reengineered it slightly and replaced the 2-stroke engine with its own R27 single-cylinder 4-stroke engine, bored out to a heady 295 cc, producing 13 bhp at 5200 rpm.

A 4-speed motorcycle transmission was used, with reverse added. The car had a short ladder frame with leading-link independent front suspension and quarter-elliptic leaf springs at the rear, with a solid rear axle/chain case bolted to the springs. The wheels were placed close together, so no differential was deemed necessary.

The body was of sheet steel, wrapped around a framework of steel tube, spot-welded together for a surprisingly rigid structure. The single front door swung outward, a cleverly articulated steering column swinging with it. There was no trunk, just a parcel shelf behind the 2-passenger seat. The whole package cruised along at about 45 mph and could be goaded into exceeding 50 at times.

Germans bought these cars because they needed them and because they were better than walking. Americans generally bought them because they thought they were hilariously funny, or because they were "cute" or "neat" or just interesting. Or, occasionally, even because they were better than walking.

Isettas suffered a lot of indignities on this side of the Atlantic. They were donated as booby prizes in contests, given away by car dealers as "a spare" and burned on top of homecoming bonfires. A body man who worked on Chris's car said that a girl in his high school class had one, and some kids set it on top of the school's air-conditioning unit. The car was destroyed when a wrecker operator used a sling through the open windows to lift it down, crushing the roof.

Chris said he used to drive his Isetta to work in Milwaukee, and the other mechanics would lift the back end as he tried to drive away, then drop it, which generally damaged the rubber donuts in the driveline.

What the hell? It was just a toy. Real cars were big, after all, impossible to lift—the more impossible, the better. Ah, we were a well-fed, healthy lot in those days, and the Isetta was the runt of the automotive litter.

I suppose to appreciate an Isetta, you have to spend five weeks restoring one, refurbishing and gazing upon each component. Only then do you realize the Isetta is not exactly a joke. No crude hardware-store project, it has dozens of specially built pieces and castings that are almost works of art. Look at the finely crafted steering box, the smoothly machined front hubs and brakes or the beautifully cast rear-axle housing, and you realize that all these were carefully engineered and produced by people who did absolutely the best work they knew how.

Stay up all night cleaning, painting and reassembling the elegant pedal cluster on an Isetta, and you would never want to see the car burned on top of a bonfire.

Nay, you would want to drive the thing all the way to Memphis.

And that is what Chris and I decided to do, the morning after our throttle problems in downtown Rockford.

We awoke well before noon—our first decent night's sleep in weeks—and ate a hearty breakfast without Doritos at a small cafe, swilling hot coffee and alternately looking at maps and our semi-trusty Isetta sitting in the cold parking lot.

"Where to?" Chris asked.

"South," I said, quoting U.S. Grant. I ran my finger over the pages of our road atlas and said, "Let's meander down the Mississippi River, deep into blues country. We could start over here at Hannibal, Missouri—Mark Twain's hometown—and head down to Memphis. Maybe check out Graceland and hear a good blues band on Beale Street. If Tom and Huck could make it downriver on a raft, we should be able to do it with a modern postwar conveyance such as an Isetta."

Chris looked doubtfully out the window and said, "Well, I hope it floats." Then he added, "One of my main goals on this trip—wherever we go—is to get through an entire tank of gas without having to repair the car. I could never do that with the Isetta I owned in college."

■■■■

The Isetta doesn't so much accelerate as accumulate speed, much the way a $20 U.S. savings bond accumulates interest.

■■■■

It seemed a small enough thing to ask, especially as the Isetta had only a 3.5-gallon gas tank. Yet we were less than 100 miles from home—a little more than two gallons into the trip—and we'd already made three roadside repairs. Most of these were teething problems, centered on the fuel system.

Chris had spurned the stock 22-mm Bing carburetor in favor of a great big honking 32-mm Mikuni motorcycle carburetor, "so it won't be as slow as the car I had in college." (Apparently, having an Isetta in youth leaves permanent scars.)

Unfortunately, he'd fashioned an in-take manifold from a section of Peugeot radiator hose, which pulsed at idle like the lungs of a hummingbird and blew up after about 50 miles. We'd put in a new piece of hose on the road last night, this time wrapped with electrical tape. It was holding so far, but didn't bode well.

My first stint behind the wheel of the Isetta was quite enjoyable. Except for the left-handed, backward-H shift lever, which protrudes from the left wall of the cockpit, operation of the car is fairly conventional. The clutch works easily, the drum brakes (only one drum at the rear) are reasonably powerful, helped, of course, by the mere 800-lb. weight of the car.

Steering is medium-light, with good tracking and stability and no twitchiness or quirks, and the handling of the car is remarkably good. It zips through corners (when zipping speed can be at-

tained) with good grip and very little body roll, yet the ride is excellent for such a short car. There's a little hobbyhorsing over repetitive road bumps, but it soaks up potholes and railroad tracks that would shorten your spine in, say, a 1958 Sprite.

Power? Well, the Isetta doesn't so much accelerate as accumulate speed, much the way a $20 U.S. savings bond accumulates interest. It's almost fast enough not to be a hazard in city traffic, juddering up through the gears with a sound somewhat akin to a paint mixer. On the highway it hums along quite serenely at 45 mph, sounding more like a Taylorcraft with a 65-horse Continental. The whole car, in fact, has the feeling of a self-propelled airplane cockpit, and when you look out the windshield, you keep expecting to see a prop disc glistening in the sun. Rate of climb is poor, and uphill speeds can drop to 30 mph or less.

This is a good speed range, however, because nearly everyone can pass you effortlessly on a two-lane road. If you cruised just a little faster, say at 52 mph, traffic would back up. Only a few cars failed to pass us immediately—generally large sedans driven by aged farmers or freshly permed women in Church Lady glasses. Even these, fraught with indecision and uncertainty, would eventually creep by, chins forward, death grip on the wheel. In many cases, I am sure, we were the first car they had ever passed.

We sped down scenic Highway 2, along the Rock River, through Dixon, boyhood home of Ronald Reagan, then took 78 south toward the Illinois River, flowing toward the Mississippi through the fall color on a cold, crisp day. In Canton, we passed a supermarket with the Oscar Mayer Wienermobile in the parking lot, so we naturally turned in and parked next to it, hoping to set off a kind of psychedelic disorientation in the shopping public. As we parked, a woman poked her head in the window and asked, "Where *is* everything?"

"Like what, for instance?" Chris asked, grinning.

"Where's the front end? Where do your feet go? Where's the motor and all the things that a car has?"

A man in an MG sweatshirt walked up and said, "Man, I haven't seen an Isetta since I was in high school."

An older fellow looked the car over and said something we would hear repeated a thousand times on this trip, with no variation in wording: "I'd sure hate to hit a semi head-on in that thing."

As we drove south, Chris and I discussed which car, exactly, we would choose for the express purpose of hitting a semi head-on. We both agreed it should be something more substantial than an Isetta, perhaps a Nash Metropolitan or a Fiat 500. Crash safety was not a subject to dwell upon in the Isetta. Before the trip, we had even decided against installing seatbelts, theorizing that "it might be better to be found as far as possible from the scene of the accident."

Chris confessed that his worst safety nightmare was that we would hit the grille of a 1950 Studebaker, while I worried most about the front bumper of a 1953 Cadillac with the infamous spring-loaded "Dagmars." Either scenario was grim.

Highway 100 took us down the Illinois River into darkness, across Route 67 at Beardstown, which Chris pointed out was one better than Route 66. This comment led, after many hours of driving, to plans for a new TV program called *Route 65*, in which everything would be slightly disappointing,

■■■■

Only a few cars failed to pass us immediately—generally large sedans driven by aged farmers or freshly permed women in Church Lady glasses.

■■■■

compared with the original series:

Two guys, named Tud and Budd, would cross the country in an Isetta, rather than a Corvette, searching for adventure and Truth. Waitresses who had always wanted to "get out of this dump and see the world" would decline to ride with them, preferring to wait for a better car; small-town redneck sheriffs would refuse to arrest them for anything; they would be unable to pick a fight with guys at the factory, or even get a job there. Nelson Riddle's cousin, Billy, would write the theme music. Worst of all, carburetor problems would prevent them from ever leaving Illinois.

At which point our latest Peugeot radiator hose/intake manifold blew out, and we coasted into the parking lot of a motel in Pittsfield.

In the morning we found a much higher-quality intake hose (GM) at O.B. Dell's auto shop—with much kind help from O.B. himself—and headed toward Hannibal, stopping for lunch at Smith's Rockport Cafe in the little village of Rockport. Best lunch of the trip.

Crossing the Mississippi bridge, we dropped down into Hannibal and pulled up near Mark Twain's house, right next to the Becky Thatcher Book Store, where I bought hardcover replacements for a few of my disintegrating Twain paperbacks. More Isetta ballast. We then drove a few miles south of town, to Mark Twain's Cave, fictitious hideout of Indian Joe, where Tom Sawyer and Becky Thatcher were lost.

Frankly, I had expected this to be a glorified hole in the wall, but it turned out to be an impressive and amazing labyrinth, miles of passages, chambers, connecting caverns and dead ends. You could easily get lost here forever, and some did. The skeletons of two local draft dodgers were found after the Civil War in a remote alcove. Many of Hannibal's children were temporarily lost in the cave, and organized search parties seem to have been a regular social function of the town.

Having grown up in a sleepy Midwestern town on a river, I identified closely with Tom Sawyer. I read the book at an impressionable age and spent one good summer of my childhood building rafts, fishing, smoking a pipe, swearing and sneaking out of the house at night. The one shortcoming of my hometown was that the only local cave was so small you could see the back of it from the highway, and so could your parents. Some hideout. So the vastness of the Mark Twain Cave was both a relief and a fulfillment.

Jesse James and his gang are also said to have hidden out in the Mark Twain Cave. But then, the James gang is reputed to have hidden out virtually everywhere in Missouri but the Lone Eagle Pub at the St. Louis airport, so a little skepticism may be in order.

From Hannibal we roller-coastered down Highway 79, the river road to Wentzville, where we found a motel. In the morning, we took small back roads south, stopping for a reverent moment at the front gates of Berry Park, Chuck Berry's rambling country estate near Wentzville. When you've spent as much time as I have playing "Carol" and "Little Queenie" on the guitar, you have to pay homage.

Meandering toward the Missouri River, we ran across Daniel Boone's last home in a remote valley near Defi-

ance. A lovely old stone and timber two-story house that Boone built by hand in his old age. Unfortunately, the simple, gracious homestead is now cluttered with a restaurant, a gift shop and the construction of a Living History Village. Poor Boone. Elbow room, indeed.

Some of the best sports-car roads in the world (if you had a sports car) led us into the wooded hills at the edge of the Ozarks, and we stopped for the night at Piedmont, after putting our muffler back on a few times and rejetting the carburetor with leaner jets from a Honda shop. In cruising through dozens of small towns, we noticed that reaction to the Isetta was nearly always the same, occurring in predictable stages: (1) accidental eye contact; (2) the classic theatrical double take; (3) "Isetta paralysis," in which kids stop dribbling basketballs, jaws drop, fuel fillers overflow and the thread of street conversation is lost in mid-sentence; (4) involuntary, reflexive pointing, where the arm comes up like a spring; (5) grinning; (6) calling to friends, frantically, to come and see the car before it's gone.

Everyone—and there were no exceptions on the entire trip—seemed to like the Isetta and thought it looked like a lot of fun, doubts about head-on collisions aside. A very common reaction was to ask a few questions, quietly examine the Isetta and then conclude, "This is all you really need to get around in."

South of Piedmont, our fifth day on the road, the weather warmed up for the first time. The sun came out and we were driving through bright fall color, side windows open. I'm told the Isetta can be a hot little greenhouse in summer sunlight, but that had not been a problem for us, so far.

The under-seat heat duct, which receives forced air from the engine fan via the shrouded cylinder fins, had proved frankly disappointing in the cold autumn weather. Apparently, most of the air molecules missed all contact with the warm engine and slipped through the fins untainted by thermal radiation. The result was something closer to refrigeration than heat, so we were able to travel without

■■■■

Chris and I toured the Graceland mansion, which caused me to revive a line I used in a story long ago: "If bad taste were dynamite, you wouldn't want to smoke in here."

■■■■

the usual Coleman cooler. Our sandwiches kept for days, and we ourselves didn't seem to age much, either.

Outside of Poplar Bluff, we hit a flat section of road and decided to do some timed acceleration runs. The results were impressive: 0–40 in 23 seconds; 0–45 in 32 sec.; 0–50 in 1 minute flat; 0–52 in 6 minutes. Fuel mileage had ranged in the mid-to-high 40s. Chris thought both acceleration and mileage would have been better with a stock carburetor.

Just north of Clarkton, we dropped down onto the flat, cotton-heavy plains of the Mississippi valley, and suddenly we were in the South. Delta country: warm sunny short-sleeve territory, with abandoned sharecropper shacks along the road, out of the old 40-acres-and-a-mule hill country and into the land of plantations, crop-dusters and King Cotton. We'd crossed an invisible line of demarcation; Deep South, northern edge.

After a hard afternoon of driving,

we cruised our little egg into Memphis, heading straight downtown and checking into the famous and ornate old Peabody Hotel, just a few blocks from Beale Street. We cleaned up and headed out for drinks, dinner and some music.

Beale Street is a famous old nightclub district that fell into disrepair years ago, but has been revitalized. I was here 14 years ago on a motorcycle trip, and the area was mostly bombed out and boarded up, except for a few bail-bond offices and liquor stores. Now it has dozens of blues clubs and restaurants, such as B.B. King's Club. B.B. himself got his start here in the late Forties, as "Blues Boy King," with his own radio show.

Chris and I ended up at the Rum Boogie Cafe, eating gumbo and listening to a superb guitar player named Preston Shannon sit in with the Rum Boogie Band. Shannon, at one point, asked people in the audience where they were from, and about two-thirds were from Europe, New Zealand, Japan and Australia. Memphis is a city of pilgrimage.

So in the morning we "pilgrimed" out to Graceland, Elvis' famous home. I had been here too, 14 years ago. By chance, I had arrived exactly one year after Elvis' supposed death, and the place was thronged with fans who lined up to parade past the graves of Elvis and his mother. The house was closed to tourism then, as family members were still living there. Now the place is a full-fledged tourist mecca, highly organized with tour guides and a huge museum across the street.

Chris and I toured the Graceland mansion, which caused me to revive a line I used in a story long ago: "If bad taste were dynamite, you wouldn't want to smoke in here." Not quite fair, of course, as none of us would care to have our tastes frozen back in the Seventies for all to see and ponder. And, as my brother, Brian, pointed out, "Elvis' strength was not really interior decoration, but he was a better rock 'n' roll artist than most."

Indeed. You read the names off the gold records on his wall, and you realize that the kid from Tupelo owns an awful lot of real estate in your brain.

Chris and I visited the Elvis car collection across the street, which, except for the staid conservatism of a few pink Cadillacs, was fairly glitzy. Elvis liked odd vehicles—3-wheelers, dune buggies and the like—with fiberglass bodywork and lots of metal flake. Given Elvis' taste for the unusual, we decided that he probably would have liked the Isetta. Had we parked it outside the Graceland gates when Elvis was still living at home, he might even have come outside to see it up close and maybe even have bought it from us, paying in cash. Then we could have flown home, avoiding Illinois.

As it was, we had to drive. And drive we did, straight as an arrow into the North, dialing summer back into a late, frozen fall with our small space capsule. Filling up as we left Memphis, Chris pointed out that we had actually used an entire tank of gas without fixing the car. A few miles down the road, our rubber intake manifold blew out and we had to adjust the points and retighten the muffler clamps, but that didn't matter. A record had been set.

We made it home in three days, with stops in Cairo and Kewanee, Illinois. A phantom engine miss plagued us all the way home, but the car kept running.

When Chris dropped me off at my house, we pulled into the driveway with the starter/generator brushes failing, the manifold collapsing and a bad sparkplug miss, but we made it.

Since then, Chris has taken the Isetta to his car shop and installed the old, original carburetor. The car goes faster now and gets better fuel mileage. He also found a short in the wiring and recrimped the muffler connection. If we made the trip again, it might be trouble-free. But neither of us has suggested it.

S till, I wouldn't hesitate to go anywhere in the Isetta again. It's quite comfortable, even for long hours behind the wheel; and long-distance travel at times has a best-seat-in-the-house charm about it, as though you are seeing America from an armchair in a bubble.

It's slow, but once you are resigned to that slowness, there's a humane aspect to the pace; no one surprises you, and you surprise no one on the road. Even the most dimwitted birds miss your windshield, farmers checking their mail along the highway regard you benignly, and their dogs and cats are safe. At 45 mph you can always stop, right now. Everything is avoidable; all accidents waiting to happen are still waiting. Probably forever.

Another part of the Isetta's charm, I suppose, is symbolic. One look at the car, with its severe economy of design, and you sense that it embodies a kind of lost freedom. No civilized nation would now permit an engineer to build a car this light, this simple, this eccentric or this logical for its specific purpose. It would be impossible to build and sell an Isetta today, without doubling its weight and bulking it out with bumpers, airbags, etc.

In any case, the Isetta makes a nice talisman from another era. It's like a Burgundy bottled before the spread of phylloxera in the 1870s; for better or for worse, we can't reproduce it and we'll probably never see its like again.

Not that anyone would want to build an Isetta again, unless resources and money should somehow become as scarce as they were in postwar Italy and Germany. We have 58-mpg economy cars now that cruise comfortably and quietly at 80 mph, carry four people and have heaters that work, so it would take catastrophically lean times, indeed, to bring us back to a design as Spartan as the Isetta's.

But then, the car was a product of catastrophically lean times. As Chris and I were reminded just before we began the Isetta's restoration.

Chris called an old friend named Klaus, in Germany, to see if he could find us a genuine Isetta shop manual. Chris and Klaus used to work together as mechanics, years ago, at a sports-car shop in California. When Klaus heard what we were doing, he said, "Oh no, Chris. Please don't tell me you have bought an Isetta."

"Why not?"

There was a long silence, and then Klaus said, "Americans think these cars are fun, but for Germans they are a reminder of very bad times, when we had nothing. We don't even like to see them."

He sent us a shop manual, nonetheless, and we were able to finish the car and take it for a drive down the Mississippi, into blues country.

For that purpose, the Isetta was something of a failure. We never did find a single person with the blues. We spent a solid week, driving all the way down to Memphis and back, and never saw anyone who wasn't smiling.

Still, I couldn't help thinking about Klaus's remarks during the trip. Would Americans have been so overjoyed with the Isetta if their own towns and garages were full of them? I wouldn't think so. In America as in Germany, it was the Isetta's rarity that created happiness and contentment. And vice versa. Cultivating the blues is a luxury, while living with them is something else entirely.

TRIUMPHS AND OTHER DISASTERS

Ah well, I *finally* had a sports car...

BY PETER EGAN
ILLUSTRATION BY HECTOR LUIS BERGANDI

"AMERICA...YOU MADE me want to be a saint" was the way Allen Ginsberg wrote it in his amazing *Howl and Other Poems*. I read those words when I was in high school, but I don't think I really understood them. It finally sank in two years later, when I was a sophomore in college and bought my first car, a 1957 TR-3A. As it turns out, it was not America that almost made me a saint, but England. By way of Triumph.

Let me explain.

In 1967 my roommate, Pat Donnelly, and I were both attending the University of Wisconsin in Madison and living in a dormitory called Sullivan Hall, an Eisenhower-era structure of cement block and glass, built in a style that Tom Wolfe would later characterize as "East German worker housing."

Pat and I were both working nights at the Coca-Cola bottling plant, unloading trucks of empty bottles, to earn money for Triumphs.

Triumph motorcycles, that is, not cars, at least that's how it started out.

The plan was, we would buy a pair of Triumphs, ride them around during the summer and then take a tour of Canada in the fall, before school started again. By midwinter, we'd each

> *British cars that run on three cylinders have an unsatisfying engine note, like a Spad going down over Flanders.*

saved around $400 toward our bikes.

Then I met Barb, to whom I would later be married.

Suddenly a sports car—which could be driven in winter—seemed like a better idea than a motorcycle.

As I already had a Triumph Fund in progress, I simply switched the quest to one of my favorite sports cars, the lovely, rugged TR-3: low doors, wire wheels, impractical but romantic side curtains and a wonderful, growly 2-liter engine with two SU carburetors.

When I told Pat I was thinking of buying a Triumph car rather than a motorcycle, he confessed that he had been thinking about buying an MGA, so he could visit his girlfriend, Maria, who was going to school in LaCrosse. Suddenly we were both in the sports-car market.

My TR-3 showed up first. I saw an ad in the Milwaukee papers for a 1957 British Racing Green TR-3A, in good shape, for $500. I called the owner and learned he was a seminary student who was about to become a priest.

I suppose in my mind I thought that he was casting off the things of this world and taking a vow of poverty. It didn't occur to me until much later that most priests don't have to take a vow of poverty. He was probably just

taking a vow of common sense and buying himself a Chevy. He may also have been taking vows of warmth, dryness or punctuality; I'll never know.

At any rate, he was a nice fellow and told me everything he knew about the car. It had new British Racing Green paint, a white top and wire wheels, but ran on only three cylinders. A garage had told him it needed a valve job, otherwise he would be asking more money. I offered him $400 and he said to come and take a look at it.

So one raw Thursday afternoon in

March, I got out of class, walked to the Badger Bus Depot in light snow flurries and took a bus from Madison to Milwaukee, where I stayed overnight with my newly married sister, Barbara, and her husband, Russ Card, in the suburb of Shorewood.

On Saturday morning, Russ drove me to the apartment building of the seminarian. I could see the green TR-3 in the parking lot from two blocks away, and by the time we got there, I'd already decided to buy it. The car looked beautiful and, as promised, ran on most of three cylinders. I shelled out $450 and drove away.

Or kind of stuttered away. British cars that run on three cylinders have an unsatisfying engine note, like a Spad going down over Flanders. That dead cylinder also works as a kind of compression brake, so I had trouble keeping up with Russ's Lincoln Continental on the way back to his house.

Nevertheless, I was driving a real sports car—my first—with fluttering side curtains, a 4-speed gearbox, roaring exhaust, damp carpets, door latches operated by a pull-cord, a bonnet opened with a Dzus fastener tool and a big steering wheel with wire spokes. I had chucked the conventional for the sublime. What a guy.

When we got to Russ's house, my sister had departed for a weekend teachers' conference and had left us a huge container of spaghetti for dinner. Also the phone was ringing. It was Pat Donnelly, calling from the Milwaukee bus station.

He had broken up with Maria after a large argument and was dropping out of school to join the Army. Could we put him up for the night and drive him to the Milwaukee Induction Center in the morning?

"Sure," Russ said. "We'll be right there."

On the way to the bus station, Russ said, "We have to keep him from joining the Army until he comes to his senses. We'll probably have to get him drunk and lock him in the closet or something."

We collected the downcast Pat at the station, and on the way home we stopped at a liquor store and picked up two 1-gallon jugs of cheap red wine with somebody's smiling portrait on the label. Obviously a "before" photo.

The wine went down very easily with all that spaghetti and the heightened emotional backdrop of temporary bachelorhood, car celebration and low-down, mean-woman induction-center blues.

About midnight, it occurred to Russ that his old pal and nearby neighbor, Bill Gether, should look at the TR-3 and see what was wrong with it. Bill owned a gas station and built stock-car engines. He did not have a high opinion of cars that hadn't been built by the Ford Motor Company, Russ explained, but he was a good mechanic. So we piled into the TR-3 and roared down back alleys to Bill's house. We brought a case of beer as a goodwill gesture.

Bill adjusted the valves, set the points and checked the timing, but still the TR-3 ran on three cylinders. By that time, the beer was gone and our diagnostic faculties had lost their keen edge. Pat was lying on a creeper near the workbench, moaning incoherently, speaking in tongues. Thick ones. So Russ and I put the TR-3's top down and slung him into the diminutive rear parcel bench and drove back to the house in the cold March air.

Pat didn't join the Army the next day. None of us did. It's hard to imagine double-timing through Kentucky with a pack and a rifle when you are too hung over to eat toast. Pat looked like Death. Or Keith Richards, only more so.

At noon we said goodbye to Russ and left for Elroy, the small Wisconsin town where Pat and I had grown up. It was a 150-mile trip and the car still gargled along on three cylinders, but the weather was suddenly beautiful. It was the first real day of spring and it had that Easter Sunday feel about it, even though it wasn't. We left the top down—keeping the heater on—and motored along on secondary roads.

In the Baraboo hills, I heard a metallic clatter and looked in the mirror to see my flip-up Le Mans-style fuel filler tumbling off the road into the steep woods below. We searched for two hours and never found it, so I stuffed a rag in the fuel-filler neck.

Arriving in Elroy with three-cylinder stress headaches, we stopped on Main Street in front of my dad's newspaper office. He came out in his apron and looked the car over, nodding. "Looks pretty good," he said, as cheerfully as a father can when he has just learned that his son in college has given a hostage to fortune in the shape of a 10-year-old British sports car with a shop rag for a gas cap. "Better not smoke," he said.

That afternoon, in my parents' garage, I fiddled with the engine and discovered it had a dead sparkplug. With a working plug installed, it fired up and ran like a new car. I went

howling off into the countryside, revving the engine and reveling in its reborn smoothness and power.

I picked up Pat at his house and gave him a demonstration ride. I threw it hard around a tight corner and the right front tire went flat.

That's when I discovered that all the spokes were loose, and hard cornering drove them into the aged tubes.

Self-puncturing tires. The new safety discovery that saves teen lives by cooling their ardor for speed.

I had two more flats that weekend while demonstrating my new sports car to family and friends. While I was fixing the third one, the car slipped off its jack and rolled over the removable body panel that covered the spare tire compartment, also breaking the Dzus wrench in half.

On my drive back to Madison, the taillights went out, and I discovered they were screwed into wooden blocks inside the rear fenderwells because the taillight nacelles were sculpted entirely of Bondo. Bad ground.

Ah well, I had a sports car, even if I couldn't drive around corners fast or count on the lights.

Back at college, Barb was suitably impressed by my stunning new car, and I invited her to the Sullivan Hall Spring Formal. She accepted and bought herself a beautiful white dress and gloves.

On the evening of the formal, all the other poor peons from the dorm had to load their girlfriends unceremoniously into a yellow school bus for the ride to the dance. But not your extremely suave and worldly P. Egan, Journalism Major.

While the people on the bus watched enviously (I imagined), I helped my beautiful Barbara into the glamorous green TR-3, folding her gown carefully into the footwell and closing the door.

On the way to the dance I honked at a car on East Washington Avenue and the Triumph's wiring harness began to melt. By the time Barb could say, "Do you smell something hot?" thick clouds of black smoke were pouring into the cockpit and the underside of the dash was on fire. I pulled over and we leaped out.

When I finally got the battery disconnected, the entire wiring loom had sizzled and melted into a charred, stinking mass of cotton, copper and plastic. The fire died down, but smoke rose from the engine compartment like an atomic mushroom cloud in a cheap cartoon.

Just then the bus from the dormitory went by. Everyone waved.

I pushed the car to the corner of a vacant lot and called a cab. Barb's dress was not damaged, though my own suit would require several dry cleanings before it lost the stigma of a fire-sale garment. We had a good time at the dance anyway and rode home on the bus. We passed the TR-3 and it looked forlorn under the streetlight, like an Edward Hopper painting.

It took four afternoons to rewire the car and get it running again. The fire was caused by the unfused horn wire chafing against the end of the steering column. A dead short. I retraced every wire, one at a time, and replaced it with a new one.

Once the car was running, I ran afoul of another problem. In midyear, there was nowhere left to park on campus. The UW discouraged students from bringing their cars to college anyway, and the campus police could ticket you for driving on campus during weekday class hours.

So I parked my car—along with 20 or 30 other renegade car owners—at an abandoned Shell station on University Avenue. The advantage of the Shell station was that your car was perpetually boxed in by other cars, so you were not tempted to drive around and waste money on gasoline.

Nevertheless, I would sometimes visit the car, just to see it. I also got a glimpse of the Triumph every night when I rode the bus out to the Coke plant, where I was still working so I could pay for car insurance.

One Friday afternoon I visited the Shell station and half the cars were gone. Those that remained each had a $25 parking ticket under one wiper, along with a warning they were soon to be towed. Also my car had a flat.

I changed the tire and decided to drive home for the weekend and leave my car there. On the drive back to the dorm to pick up my suitcase, the campus police stopped me for illegally driving on campus and gave me a $30 ticket.

I now had tickets equal to the value of a month's pay at the Coke plant.

Nights and nights of unloading bottles, gone.

On the way home to Elroy, I hit a bump and all my lights went out. Bad ground in the vicinity of the wood-and-Bondo taillight zone.

During Easter vacation, I decided to drive to Wausau, Wisconsin, to see

Barb and meet her parents. I put on tan slacks and a sweater, so as not to look like the student radical lowlife I really was, and headed north.

The TR-3 broke down three times on the way to Wausau. Broken fan belt, two blown radiator hoses. The

---

My dream car was keeping me broke and making me crazy. I had become its full-time servant, a worker bee serving its queen.

---

last radiator hose blew out just a few miles from Wausau and green coolant came lurching out around the hood, streaming over the windshield. I wrapped the hose with black plastic electrical tape and was in the process of toting green ditch water to the radiator in a rusty beer can, 12 oz. at a time, when a new Cadillac stopped and a dignified-looking gentleman in a gray business suit got out. He watched me for a minute, then found his own rusty beer can and began to help fill the radiator with green ditch water,

getting his black wingtips wet.

"You must be Peter," he said. "I'm Barbara's father. I thought this would probably be your car."

Barb's dad spent most of his weekend driving me all over Wausau to find replacement hoses that would fit a TR-3. We finally found some at a Wausau Mercedes shop; then Barb and I drove all the way to the top of Rib Mountain and back to the house—a trip of more than 12 miles, without incident. This was more like it.

Barb and I drove back to college together that Sunday in a pouring rain. The air vents leaked and flooded our feet with rainwater as if they'd been designed by clever hydraulic engineers to do just that, and the side curtains bowed out in the wind like a pair of chin scoops on a P-51, giving us ram-effect rain in the cockpit. I stopped on the Interstate to get some rags out of the trunk and a semi went by. The wind blast snapped off both my pot-metal trunk hinges. By the time we got to Madison, Barb's clothes were pretty much ruined, but she never complained. England was making her a saint too.

I was getting further and further behind the repair and finance curve on the Triumph, but that didn't keep me from taking a weekend trip to see my old pal Jim Wargula, in Oshkosh. I got there with no problem, but when I took Jim for a ride we went over some railroad tracks and the exhaust system fell off the Triumph. We retrieved it and spent the weekend patching a gaping muffler hole with fiberglass tape and bolting the system back onto the car.

While we were picking up parts, we ran around town in Jim's maroon 1966 Mustang. It was almost unearthly being inside the Mustang after a 100-mile trip in my Triumph. There was no wind noise, and the heater and defroster worked. It was raining out, yet no rain came inside the car. Not only did the Mustang have a radio, but you could hear it. Jim drove fast around corners, yet had no flat tires. The suspension moved up and down in response to potholes.

Maybe I'd been too hard on American cars.

The fact was, there was nothing wrong with the Triumph (well, almost nothing) that couldn't have been cured the first week I owned it, if I'd only had a little knowledge and extra money.

If I bought the same car now, I'd bring it home and change the belts and hoses, do a tuneup, buy some new tires, a new exhaust system and tighten the spokes and true the wheels. In fact, I probably wouldn't have even bought this particular Triumph, because I would have spotted the wood blocks and the Bondo in the rear end.

But in 1967 I was trying to operate a car on a bicycle owner's budget. It was hopeless. The failure of any major part was a financial disaster that could wipe me out for a month or more. The car—my dream car—was keeping me broke and making me crazy. I had become its full-time servant, a worker bee serving its queen.

The last straw came on a spring afternoon during final exams. The car was now parked on a side street near campus—I had to keep moving it every 24 hours to avoid tickets—and I went to move it before class. I cranked the engine and the starter went up in smoke. I went back to the dorm to get some tools, and when I got back, the car had been ticketed.

I threw the tools in the trunk, turned away from the Triumph and began walking.

All afternoon, in a daze.

I walked out to Picnic Point, then along the Lakeshore Path and down to the student union building. I sat on the terrace, smoked cigarettes, drank coffee and looked at the lake.

It was all a joke and a sham, this material thing. Dreams based on objects came to nothing; they slipped and sifted through our fingers like sand. We were, like James Joyce's autobiographical Dublin boy, nothing but poor creatures "driven and derided by vanity." Property was not a crime, as the New Left was telling us; it was slavery. Tolstoy and Thoreau saw it a long time ago, even before Triumphs were popular.

I concocted throughout that afternoon on the terrace various plans to divest myself of material goods and simplify my life. Should I join the Peace Corps? Become a Trappist? Perhaps a Franciscan order...

As the day wore on and the sun slanted low on white clouds gathered over the lake, I began to calm down. No need to go overboard. I was teeter-ing on one of those philosophical brinks that can lead one so easily to toss in the cards and go looking for a whole new game. Dangerous stuff.

What I would do is sell the TR-3, and think about the rest later.

I had no luck selling the car in Madison. No one wanted a Triumph with a fried starter motor, and of course I didn't have the money to fix it.

"Push-start that thing and drive it to Milwaukee," my brother-in-law Russ said. "You'll be more likely to sell it in a bigger city. I'll find a buyer for it."

So I dropped the TR-3 off in Milwaukee, and Russ put me on the bus back to Madison. Just the reverse of my original trip. A few weeks later, Russ sold the car to some uncritical, starry-eyed soul and sent me a check for $400.

The Triumph was gone. No more tickets, repairs, fires, flats or insurance payments. Free at last.

I bought a used Honda CB-160 motorcycle for $200. The bike ran flawlessly and averaged about 80 mpg. I could also park it almost anywhere. That fall, Russ lent Pat his old 305 Honda, and we took our motorcycle trip through Canada.

As it turned out, motorcycle ownership was just the right level of self-denial and sainthood for a mechanically minded person in my income bracket. I never did join a monastery or a commune or give away all my material possessions.

I did, however, quit school that next winter and join the Army. I went in almost a year from the day I'd bought the TR-3. Russ picked me up at the bus station and drove me to the Milwaukee Induction Center early in the morning on March 4, 1968. He was also there, with the rest of my family, to pick me up when I came home from Vietnam in 1970.

By the time I got out of the Army, I'd grown up a lot and learned a few lessons. I bought another TR-3 when I got out, but not until Barb and I had pooled our resources and bought a clean, reliable, low-mileage 1968 Volkswagen Beetle as our main car for actual transportation.

We used the VW and a piece of rope to tow the new Triumph back to our house the day before we were married. We got the car cheap because it had a bad starter.

Russ lent us his Lincoln Continental for our honeymoon. He didn't have to, but all those trips to the bus station had made him a saint. ◎

138

ASKING A JOURNALIST who lives in Wisconsin if he would like to fly to Sweden in late December and drive a Volvo 850 GLT on a frozen lake north of the Arctic Circle is about like asking the martyred St. Sebastian if he would like one more arrow to pierce his flesh. "Or is that quite enough arrows for now, sir?"

It was an exceptionally dark winter in the Midwest this year, so the offer was not quite as seductive as, say, a comparison test of luau catering trucks in Maui. But I readily agreed to make the trip anyway, as Sweden is one of my favorite places, regardless of its meteorological similarity to my home state. Why?

Well, Swedes have been criticized for being rational and fair-minded to a fault, but I have found this generalization to be overstated. Many Swedes really are rational and fair-minded, but they also know how to have a good time, tell jokes and tip a few. As a result, Sweden seems to be one of the last really civilized places on earth. Good manners and good cheer prevail.

I hope it stays that way, because these are hard times for this country of 8.5 million thinly spread souls. Their economy, like that of the U.S. and much of Europe, is in tough shape, with bank bailouts, currency problems and the usual "peace dividend" cutbacks in their sizable defense and aircraft industries.

Philosophically, the country is also somewhat adrift. Sweden has always stood for the "middle way" between Soviet communism and Western capitalism, but with the breakup of the old U.S.S.R., it is in the position of being on a bridge that goes nowhere, now that the Evil Empire on the other side simply collapsed.

To reach Sweden, I flew to Newark, met a bunch of other journalists and Tomas Ericson, Volvo's man in the U.S., and we all boarded an SAS flight to Stockholm. There we transferred to a smaller plane and took off for the north country, flying over an ever-snowier landscape of lakes, farms and wooded hills until we reached Kiruna, the northernmost city of any size in Sweden, only 1500 miles from the North Pole.

Kiruna is a large iron-mining town of 30,000, but it has a small windswept

IN THE LAND OF THE MOONLIT AFTERNOON

Our Editor-at-Large wrings out Volvo's new front-driver in Sweden's reindeer country

BY PETER EGAN
ILLUSTRATION BY DENNIS BROWN

airport that looks like something from a John le Carré novel, the kind of place where Smiley and Karla might exchange microfilm. We landed at 12:35 p.m.; and the sun was shining weakly, about two inches off the horizon, getting ready to go down again, having risen at about 10:30 a.m. Daylight, such as it is, lasts about four hours.

Theoretically, this should be depressing, but it isn't. With the sun just under the horizon before and after daylight, there's a luminous cast to the sky, causing the snow-covered evergreens and hills to radiate a kind of phosphorescent purple glow. Rather than being merely dark, the whole landscape takes on an enchanted, magical quality, as if lighted from within.

Odd, but beautiful.

We were bused to the historic Lapp village (the Lapps call themselves Sami) of Jukkasjärvi, about 10 miles east of Kiruna, where we checked into tourist cabins at a resort called the Jukkasjärvi Wärdshus och Hembygdagärd, which can be pronounced only by Nordic people who drink glögg, a mixture of hot spiced wine, aquavit, almonds and raisins, which is served everywhere on the slightest pretext in northern Sweden.

The resort is right on a frozen lake and owned by a lively entrepreneur named Yngve Bergqvist, who is famous for building the "World Largest Igloo" every winter. He uses huge, cathedral-like plywood forms; heaps snow on them with bulldozers; packs it down and then removes the wood, leaving a kind of four-winged snow temple. Inside is a bar made from blocks of clear ice, backlit to glow a surreal Aqua-Velva blue. Drinks are served to the patrons, who have been issued snowmobile suits, boots and fur hats at the resort office.

At the edge of the lake was a more traditional structure, a Sami *Käta*, a pole and canvas affair that looks almost like a Plains Indian teepee. Inside was a central fire ring, where a couple of guys were frying marinated reindeer meat in a Paul Bunyan-size pan, served with glögg, of course.

Our little group spent a wonderful evening in these two structures, alternately warming ourselves and going out to stand on the frozen lake in the clear, cold night to watch a full eclipse of the moon in progress.

Later, a bunch of us took a tour of the village on some little sledlike vehicles called *sparks* (pronounced "Schpark," meaning "jump"). A *spark* is nothing but a minimal chair attached to a set of flexible steel runners, with a

Continued on page 152

The Mille Miglia and

A thousand miles in the footsteps of giants

BY PETER EGAN
PHOTOS BY RICHARD M. BARON

the Great Yellow Beast

As WE HOWLED into the Piazza del Campo in the ancient walled city of Siena, a great cheer went up from the assembled multitudes who had packed themselves into the *centro storico*, the historic center, of the town.

Police held the crowd back, and a path was cleared around the square to a platform with banners flying, where local officials and race queens waited, giving out handshakes, warm greetings and small bags of local produce.

Lunch for the drivers—cheese, bread, salami and mineral water from the nearby hills, and maybe a small bottle of wine to be left undrunk until later. Our stuff is the best; we eat well here. Good luck! *Avanti!*

Italian fast food.

Gil stuffed the gear lever into 1st, showered the crowd with earsplitting 4.1-liter V-12 Ferrari revs and slithered out of the Piazza, down cobbled medieval streets with flags and people hanging out of windows, waving and shouting "Ferrari!"

We blasted through a narrow stone arch, built centuries ago to accommodate an ox cart with solid wood wheels, out into the newer part of town where still more immaculately uniformed troops held back traffic and urged us through, emphatically. We were already going about 70 mph, but the cops were waving their traffic wands impatiently, like park rangers trying to get an elderly couple in a motorhome out onto the highway. "Faster! Go!"

Gil hit 4th, bottomed the suspension on a low crossing and shrieked up the hill and out of the city. Notching forward into 5th, he turned and looked at me, and I realized we were both grinning with that permanent death-skull grin again, the kind that gives you a headache from too much pleasure. Surely our faces would stick this way forever.

"I swear," Gil shouted over the V-12 racket and wind, "you couldn't hold this event in any other goddamn coun-

try in the whole world. It would just be impossible!"

Wonderful, sad and true, all at the same time.

Where else, indeed, would police and soldiers from a hundred jurisdictions join forces, as if of one mind, to speed your passage through an entire nation? No hold-outs, no "Let's teach 'em a lesson when they come through Plainville, boys." Not here.

Where else would teachers lead their grade-schoolers, en masse, to a stone parapet overlooking the road, so the children might wave red-arrowed Mille Miglia flags and Mercedes banners (Mercedes helped sponsor and promote the race this year) at race cars rocketing through their village?

Where else would whole farm families of three generations set up tables, chairs and a small feast under the shade of a tree along the road to cheer on fast-moving automobiles, reveling in the dust and sound?

Where else would hunched old women in black stand at garden gates to wave white handkerchiefs at automobiles. Or a shrunken, aged, one-legged man stand on crutches along the route with tears in his eyes, gesturing with one crooked finger to an ancient lapel pin on a suit coat?

A former finisher? Did that old man drive a Fiat the first year Biondetti

PILOTI: GIL NICKEL
PETER EGAN

won? An Alfa crew member for Nuvolari? What was he thinking?

It brought tears to your own eyes to see it. These people remembered. They remembered that other world, an earlier time when any excitement was possible and the naysayers of this earth were considered shrill and small, not worth your time.

They remembered an age of Titans, and they were happy to see us driving the cars that had belonged to the Varzis and Ascaris and Taruffis, bringing them back into the sunlight.

Two and a half days of driving a fly-yellow 1951 Ferrari 340 America Vignale Spider over a thousand miles of Italian roads in the modern road rally version of the old Mille Miglia made you wish for huge long arms so you could get them around this country and pat it on the back. What spirit, unaffected joy, color and beauty.

There's no going back after you have driven the Mille Miglia. Your life is changed forever. I guess I knew that when Gil Nickel called and asked if I'd like to co-drive his Ferrari.

I looked around my office that day and saw my old motorcycle and car racing helmets, sitting on the shelf as bookends. I had saved them all these years for…what? And next to them on the shelf, a pair of stringback driving gloves bought years ago at a little shop in Maranello and never yet worn. Illegal for modern racing. Not fireproof.

It seemed I'd been keeping all this old traditional stuff for a lifetime, just on the chance it might be needed for some special event. And here it was.

"Yes, Gil, I'd love to co-drive," I said.

Gil, whose own vintage helmets had been lost in shipping to Italy, asked if I would bring mine along. Perfect. He

■ **Entering the town of Radicafoni in a Mercedes-Benz SSK. An Alfa Romeo at scrutineering in Brescia's Piazza della Vittoria.**

could wear the old Moss-like Everoak and I would wear my 1966 Bell TX500, the first helmet in which I ever raced a sports car. And the string-back gloves would finally get to grip a proper wood-rimmed steering wheel.

My friend Gil Nickel owns Far Niente winery in California and is a fast, competitive vintage racer. He's been racing for years in the U.S., but this season he and his companion, Beth Yorman, decided to spend a summer living in Italy and vintage racing in Europe.

Packing light, as usual, they took along the Ferrari 340 America, a Lotus 23B, a Lotus 27 Formula Junior, a 1958 Alfa Giulietta Spider, for street use; a BMW R1100RS motorcycle; and a stunning 1929 Belle Isle Bearcat cigarette boat named *Miss Liberty*, for

■ **Stirling Moss, 40 years later, in his Mille Miglia-winning Mercedes-Benz 300SLR. Crowds (below right) in Siena's Piazza del Campo.**

lake use. Not to mention a race transporter manned by mechanics Chris
Chesbrough and John Van Trest, a fax
machine and many cases of Far Niente.

In other words, they did exactly
what I would do if I'd thought to in-
vest in a vineyard instead of a lawn.

They found a nice place to rent, the
downstairs of a house belonging to a
young couple, Danilo and Petra Gerletti, in the lakeside village of Ossuc-
cio. Like me, they took Italian classes
last winter, the better to integrate.

The day I flew into Milan, Gil
picked me up in his Alfa Spider, and
we resonated our way out of the industrial, truck-rich Po Valley toward the
white-capped Alps.

Lovely lake, Como. Shaped like an
upside down Y, with a good mixture of
tourism and real villages, royal villas
and modest fishermen's houses. The
mountain shores knife steeply down
into the lake, so the red tile roofs are
stacked and tilted house above house.

People have lived on Lake Como for
a long time—neolithic fishermen,
Greeks, Romans, Lombards, Franks,
Austrians and Germans in *Wehrmacht*
gray. Mussolini and his mistress,
Claretta Petacci, were executed against
a wall about a mile from the Gerletti's
house. There is a small plaque.

Danilo's mother, Angelina, remem-
bers sitting on her front porch looking
down at the tops of miles of German
trucks headed for the Bernina Pass as
the Americans approached. *"Ah, i
tedeschi,"* she says with a subtle Italian
hand gesture one might use absent-
mindedly to shoo a fly from the page
of a good book.

Across the lake is the setting for
Alessandro Manzoni's *I promessi sposi,*
Italy's greatest and most quoted novel.

All in all, it's a fine place to keep a
Ferrari for the summer.

When I arrived, we took Gil's Ferrari out for a familiarization drive. I'd
co-driven this car, briefly, last year in a
road rally at the Monterey Historics,
but it was good to get back in it.

The Ferrari 340 America Vignale
Spider is essentially a compact, AC
Bristol-size sports car with a great big
lump of a Lampredi 4.1-liter V-12
shoveled under the hood. It's about
300 bhp in a 1980-lb. car. There was
only one made in this specific body
configuration (with three other variants
in the same style), but it was much
copied—you can see the echo of its
clean, graceful lines in a dozen later en-
velope-bodied sports cars of the Fifties.

A friend of Gil's once dubbed it
"The Great Yellow Beast," and that's a

pretty good description.

Heavy clutch, heavy gearshift, medi-
um-heavy steering (you need that big
wheel) and high braking pressure. In
handling, it's remarkably well-balanced, with easy power-on oversteer
on tap. The 340 is brutish, fast, tough,
hot and loud. Yet, like so many Ferraris, it gets lighter and finer to the
touch the faster you go. It wants to be
driven hard, all the time.

So we drove hard, down to Brescia.
A fast *autostrada* cruise and then a

slower drive through town, in search of
the famous Piazza Vittoria, where the
cars are registered. In heavy city traffic,
the 340 is no Honda Civic. In fact,
there is not one attribute of the Ferrari
that conjures up the word "civic."

Huge crowd scene in the Piazza.
Fans, fashion photographers, race offi-
cials, drivers, a mass buzz of pleasure.
There we ran into R&T Editor-in-
Chief Tom Bryant, who would be
co-driving a 1955 Mercedes 300SL
Gullwing with Fred Heiler, from Mer-

■ Bentleys (far left), a BMW 328 (above left) and a Zagato-bodied Maserati (left). Above, a Klementaski-style photo of Gil Nickel doing his Peter Collins imitation.

cedes-Benz. We also met up with R&T Editor-at-Large John Lamm and Art Director Richard Baron, who would be shooting some pictures of our cars en route.

Ahead of us at the registration booth was Stirling Moss, trying his best not to make eye contact with fans so he could get back to his legendary 300SLR and apply the number discs to the car in which he won the race in 1955. He was number 300 on the starting roster, out of 340 cars.

Ahead of us also were various Italians and Brits who had cut in line. The concept of a queue baffles most Europeans, who prefer the sneak side attack.

Gil watched this process for about half an hour, then took a Latin fellow firmly by the back of the neck, slid him backward several feet in a sort of forced moonwalk, looked him in the eye and said firmly, "Ah'm next." Somehow, his Oklahoma-tinged English was understood perfectly.

Suddenly, we were registered. Number 226, departure time 21:48:45. "About 12 minutes to 10 in the evening," I told myself aloud.

The basic principle of the modern Mille is to allow reasonably spirited driving, but to remove the need for crazy speed. If you drive briskly and efficiently to the next checkpoint, you will get there 10 or 15 minutes early. Then you can pull over (as long as you are more than 100 meters from the timing booth) and wait, theoretically timing your finish of that stage to the exact hundredth of a second. You lose one point for every hundredth you are early or late.

If you have car trouble, or take a long lunch, however, you may have to drive like hell to get there on time. A nice setup to encourage fast—but not deadly—driving.

Incidentally, the modern rally version of the Mille Miglia has been won only twice since 1977 by non-Italians.

It takes experience and a fine understanding of the rules to win. In other words, Gil and I didn't have a prayer. He's from Oklahoma and I'm Irish.

I am also, mathematically speaking, a certified moron. Gil is actually a math wizard, but like me, he prefers fast driving to splitting hairs. His motto is, "If I ever win a timed rally, I will carry the shame to my grave."

With all these strikes against us, we nevertheless contrived to do our best, rather than blow it off and drive, as some fun-seekers do.

A glistening, glamorous nighttime scene at the start. Cars lined up for miles, driving up the ramp under the camera lights to be waved off, one every 30 seconds. Here the likes of Sophia Loren or Ingrid Bergman used to throw arms around their Ferrari- or Maserati-shod mates and wish them luck. Cigarettes were tossed imperiously to the ground, gears engaged.

Gil and I don't smoke, but it was a great start. The flag dropped, the Ferrari howled, a cheer went up and we were launched into the night, headlights playing off two walls of humanity as raindrops began spattering against our low windscreen.

It was a chilly, wet night, but the Ferrari gives off so much heat you barely need a jacket. Also, wind-flow over the screen is surprisingly serene; you can almost converse in this car, with a low-level shout. The only slight inconvenience is right-hand drive, which

prevents the driver from seeing around traffic before making a pass.

To counter this blindness, Gil and I developed two succinct hand signals: a forefinger snapped straight forward, meaning "Go for it!" and a hand held flat, palm down, meaning "If you try to pass this truck now, we will never learn to play the piano."

Life or death; a simple decision. The signals worked.

More rain. We pulled over, leapt out and put on our motorcycle rain suits. Let it rain. And it did.

The first leg of the Mille is a relatively short drive, 179.7 km to Ferrara. We got there a little early, crossed on time and parked in the *centro storico,* where a dining tent had been set up to feed and water the drivers. The rain stopped and it was suddenly a nice evening, even if it was 12:38 in the morning. We cabbed to a modest hotel and went to bed.

Up at 6:30, to get ready for our 8:33:45 departure time.

The Mille Miglia is essentially like basic training in the Army: a test to see how you function under duress with almost no sleep.

Clockwise we headed around Italy, down toward the sunny Adriatic coast, through Ravenna, Pesaro and Ancona, then across the foggy, still snowcapped Apennines to Rome for another late night arrival. This would be a long day, 618.41 km.

We were doing pretty well on our timed sections until *Gil* (notice it was not *me*) accidentally entered the 100-meter no-stop zone before a stage finish line 15 minutes early, and we automatically amassed, by rough estimate, about 340,000,879,000,547 penalty points.

Gil cursed fervently, but I was secretly pleased as punch. This took the pressure off me and my wretched time-keeping skills.

When Gil was driving, I had been giving him 20-second countdowns as we crossed each stage and had discovered that I could not read one number on a stopwatch and say a different number aloud. Dyslexia.

It would not be our year to beat the Italians.

At Ancona, our stage ended in a park on the waterfront, and we got out to have a couple of cappuccinos (the fuel of champions) and to admire the other cars. Tom Bryant and Fred Heiler were there, running a fine trouble-free rally, and Tom said the Gullwing was remarkably comfortable and quiet at all speeds. I gazed into the interior

and it did look like the lap of luxury next to our hair-shirt Ferrari.

Walking back to our car, I spotted the great Olivier Gendebien, four-time Le Mans winner, sitting in another of the Gullwings Mercedes had brought together for the Mille.

I stopped and shook his hand.

He was enjoying the drive, he said, but hoped no one mistook this rally for the original Mille Miglia. "This is not a race," he said. "It's a promenade. But the original Mille Miglia, ah, that was the greatest road race in the world."

He looked up at me intently to see if I understood.

"Imagine," he said, "hundreds of cars driving absolutely as fast as they knew how to go, for 1000 miles on these roads, all in one day. And the greatest drivers in the world at the wheel. It was glorious. Everyone in Italy—hundreds of thousands of people—lining the route, cheering madly. You can't imagine…there will never be anything like it again. It was just the best…"

Gendebien knew whereof he spoke, having finished 3rd overall in that final 1957 Mille Miglia, despite being relegated by Enzo Ferrari to a 3.0-liter GT rather than one of the big 3.8- or 4.1-liter open-class sports cars.

It was Gendebien that Alfonso de Portago was trying to stay ahead of when he refused a tire change at his last pitstop. A few miles later the tire blew, causing the deadly crash that killed Portago, his navigator, Edmund Nelson, and 10 spectators at Guidizzolo, when they left the road at

PHOTO BY THE AUTHOR

an estimated 165 mph, just 25 miles from the finish. It was the crash that ended the Mille's long run, from 1927 to 1957.

Gendebien's performance in that race is still considered one of the legendary drives, and it was wonderful to see him still with us in 1995, be it rally, race or promenade.

From the sunshine of the coast we headed into the dark, wild beauty of the mountains, slithering damply through a couple of glacial passes near Mounts Vettore and Terminillo, then down some of the most beautiful mountain ridges on earth, toward Rome.

Wherever we stopped in the windswept mountains, the spectators were freezing, but we were snug and warm, nestled next to our trusty Lampredi V-12. I'm thinking of having one installed in our house this winter, inside the fireplace. Talk about a roaring fire…

The stop for the night in Rome is one of the less charming ends to a stage on the two-and-one-half-day route, essentially an *autostrada* oasis next to Rome's Anulare, or outer beltline. Nevertheless, the bed was horizontal and sleep was good, especially after Gil and I had a small sip of sacramental Jack Daniel's at the hotel bar.

A hearty four hours of sleep found us back at our car, feeling wonderful and ready to face the last 642.01 km. Gil asked me to drive, even though it was his turn. He had a mild headache and was thinking of throwing himself in front of a train to make it stop.

A long day but a stunning drive through the sweeping castle-strewn landscape of Tuscany and out of the green mountains into the plains of Emilia-Romagna, through the great cities of Viterbo, Siena and Firenze—across the Arno, left at the Ponte Vecchio, through the Piazza del Duomo

■ **Olivier Gendebien, the endurance-race genius.**

and out again—headed for the mysterious swirling mists and dripping black trees of the famous Futa and Raticosa passes.

Through Bologna for a stop at Modena, where we performed our only wrenching of the trip, a lost hour fiddling with a sticky adjuster on the front right.

Well behind, we blazed toward Parma and Montichiari with a red sun going down through the Lombardy poplars, just making our next stage with 7 seconds to spare. The last dash back into Brescia was through built-up cities and suburbia, and most of us used the elusive "center lane," lights flashing, cops helping, motorists moving aside, to make the finish.

In darkness we sped into Brescia, turned onto a tree-lined avenue and cruised down a narrow gap between two fences holding back thousands of spectators. As we headed toward the finish line and the TV camera lights, Gil turned to me and said, "We made it! I can't believe it; a thousand miles without a mechanical failure, a wrong turn or a dent on the bodywork."

As we slowed for the finish, hundreds of hands came out from the fence, as people leaned over to high-five us.

Just then, someone missed my hand and hit me right in the face. Bull's-eye. My glasses went flying, my helmet visor snapped in half, and I thought for a moment my nose was broken. I checked for blood. Nothing. I was okay. Just reeling from the blow. I took my helmet off because the broken visor was hanging in my face and found my glasses on the floor.

Just then the generator light came on and the car died. No starter.

I jumped out and pushed it for a restart, and the Ferrari fired to life. We crossed the finish line, waved to the cameras, squinted into the camera lights, and I responded with the crisp clarity of Mike Tyson after a hard 12 rounds to an interview microphone that was thrust in my face. "Wonderful drive," I said, checking with my tongue for loose teeth.

Tired, dazed, battered and happy, all at once.

The Mille Miglia was over, and Gil and I motored away from the finish line into the dark, busy streets of Brescia to look for our hotel. The car died again—and again—and the headlights quit working. We restarted, but the engine was overheating badly in the traffic, missing and coughing. An hour later, we found our hotel, without an ounce of energy left.

We had a glass of champagne, provided by the hotel bar, with some of the other drivers who'd just come in. We toasted the Ferrari, the Mille Miglia, Italy and just plain finishing.

How did people drive this race, we asked each other, all in one day, at full racing speed? Moss and Jenkinson did the whole thing in 10 hours, 7 minutes and 48 seconds in 1955. A 98.53-mph average speed.

So much to see, so many corners, so many thousands of split-second decisions. A thousand miles of back-road Italy makes you feel as though you've been to the moon and back, or traveled by time machine through three centuries of Western civilization or just returned home from a six-year Crusade. How could anyone have driven it at speed, let alone digested the meaning of that distance, in one day?

No wonder Portago declined to give up his position over a mere damaged tire, so close to the finish. He'd lived an entire second lifetime of breathtaking chance in one day and was not about to have it all ruined by mere molecules of rubber and rayon cord. He'd already cheated death a thousand times by the time he got to Mantua Control. Certainly it could be cheated once or twice more.

Also, he was tired.

Suddenly, I could better appreciate what I had read of that last race in 1957, and the years before. Maybe that's the real value of the modern Mille Miglia, beyond the obvious pleasures of camaraderie and the beauty of seeing all those great cars on the road again. We get to glimpse, as if through a door rapidly opening and closing, the light of personal courage that still gives our own age its small resonant echoes of romance and glamour, nearly four decades later.

Those old men and women along the road remember. They weren't waving to us. They were waving to the ghosts of those who once drove our cars absolutely as fast as they knew how to go. ⊙

■ **A Fiat Balilla on the road (above), while our heroes' return is documented on the *televisione*.**

B-ING THERE

Restored at last, the MGB gets its shake-down trip on the last road to Atlanta

BY PETER EGAN
PHOTOS BY THE AUTHOR & CHRIS BEEBE

DEADLINES, WHEN THEY carry no actual threat of death, are remarkably flexible. And so it was with the MGB. First the car was going to be ready for spring. Then it was the weekend of the Chicago historics...Labor Day weekend...Hank Williams' birthday...the 60th anniversary of the repeal of the Volstead Act, and so on. If this were a B-movie, pages would have been peeling off the calendar and dropping to the ground like autumn leaves. I finally settled on a drive to the Sports Car Club of America's annual Runoffs.

Why the delay?

Well, there was the usual problem of Sloth, not to mention its many Freudian handmaidens, such as fear of transmission tunnel grease, upholstery dread and wiring harness aversion.

And then there was money. Restoring even so popular and low-cost a car as an MGB is not cheap these days. And this car, you might say, was an interesting case in point.

For those who have missed the hundred or so columns I've written about

the MGB, this car is a 1973 Tourer, last full year of the desirable chrome bumpers, normal ride height and twin SU carburetors. The car spent nearly all its life in the sympathetic hands of Patti Baron, wife of R&T Art Director Richard Baron, which is how it came to have 194,000 miles on the clock.

Patti, a nurse, commuted to work in salt-free, sun-drenched Pasadena for 10 years with the MGB, so it had no rust. As in zero. And she obviously drives well, because it had no dents or Bondo and an original transmission with perfect synchros. An engine rebuild at 100,000 had kept oil pressure and compression out of the red zone.

But miles is miles. By the late Eighties, the One Horse Shay syndrome had engulfed the car; i.e., everything was wearing out at once. It needed brake hydraulics, a new radiator, top, interior, kingpins, etc. The Barons set the car aside for possible restoration and asked if they could store it in our airplane hangar at Corona airport.

So every time Barb and I pushed

our yellow J-3 Cub out of the hangar, we got a glimpse of the old dusty B parked under the wing. Apparently, I got used to the idea of these two as a matched set, because when Barb and I decided to move to the rural wilds of Wisconsin in 1990, we bought the car from Patti with the promise that we would restore, love, honor, etc., sworn on a stack of Haynes repair manuals.

At that time, the MGB more or less started and ran, but pumped a lot of steam from a leaking heater core onto the windshield. Sheepskins covered the disintegrating upholstery, and wherever we parked the car it left a neat signature of five distinct puddles: coolant, engine oil, transmission oil, differential oil and clutch fluid. Front and rear springs sagged sadly.

Against our advice, the truck driver who shipped the MGB to Wisconsin parked it on the top rack of a six-car transporter, right over the roof of a perfectly nice silver Dodge Omni. English fluids, of course, dribbled down on the Omni, coating it with a sticky green goo. By the time it reached Wisconsin, it was flocked a springtime mixture of twigs, leaves, seeds, milkweed fluff and dandelion fuzz. I described the Omni at the time as looking like "a Rose Bowl float sponsored by a natural food store."

Home at last, and time to go to work. Three years ago the green MGB rolled into my palatial garage/work-shop for some minor repairs, "just to keep it on the road and have some fun for a while," but one night I got carried away on a dangerous mixture of French roast coffee and Guinness and accidentally disassembled the whole car.

Now there was no turning back; it was full restoration or nothing. I got out my stack of British car parts catalogs and went to work, beadblasting, painting, rebuilding and ordering parts. The UPS truck came up our driveway so often the dog would lift only one eyelid.

What did the UPS truck bring?

When it was all said and done, the MGB needed—and got—a full engine rebuild, new interior and top, new springs, kingpins, bearings, brake and clutch hydraulics, brake rotors, shoes and pads, radiator core, heater core, oil cooler, clutch, alternator, two new 6-volt batteries and cables, fuel pump, fuel gauge sender, a new/used over-drive transmission, mirrors, heavy-duty anti-roll bars, grille, emblems, rubber trim, windshield, tires, Monza free-flow exhaust system and a few hundred other gaskets, seals, bearings, hoses, lines and clamps too numerous to mention without overloading the hard drive on my computer.

Total parts bill?

$7256.50.

And then there was machine work (.040-over pistons, .010-under crank, valve seats, cam bearings; valve grind) and engine compartment paint (Del-tron), both of which I farmed out. Not to mention oil, solvents, sandpaper, brake fluid and the usual miscellany.

So the total cost of my MGB, in-cluding its original $1600 purchase price, is just over $9000. Add the cost of trucking it here from California and the cost of the wheels it still needs—the original Rostyle wheels are all slightly wonky—and I'll have a very nice $10,000 1973 MGB. Assuming my labor is free, which it always is.

So, you ask, isn't that too much to spend on an MGB?

Of course. You could buy a nicely restored one for that, or less, and save yourself months and years of work.

There are really only four good ar-guments for doing a full-on restoration

yourself these days: (1) When you are all done, you know exactly what you've got; (2) it's a way of getting into a car cheaply and then making "payments" at your own pace; (3) car restoration is a fine hobby and a labor of love; and (4) if you foolishly disassemble your entire car one inspired evening, no one will give you more than $300 for all those greasy parts.

For all those reasons, the car finally

got done. Last October.

I hooked up the two 6-volt batteries, turned the key and we had cranking, oil pressure, ignition. The engine started and ran. Sort of.

I drove the car and found that my "3/4-race" camshaft, a back-shelf mys-tery item lent on speculation by my good pal Chris Beebe, made all its power above 6000 rpm (redline) and produced an idle that can most chari-

One night I got carried away on a dangerous mixture of French roast coffee and Guinness and accidentally disassembled the whole car.

One of the best days of driving I've ever had, listening to the exhaust resonate, looking up into the tunnel of fiery red and sun-yellow trees.

tably be described as "lumpy." In the same way that a mortar barrage is lumpy. Back to the stock cam.

My other brilliant tuning idea was the use of a "bolt-on" side-draft Weber carburetor, which I never could get to respond as well as the two stock SU carbs, even with a handful of jets and a week of tuning runs. Probably needed more cam.

So with just days to go before Road Atlanta, I rebuilt the stock SUs and bolted them on. Suddenly the engine ran perfectly. When you get right down to it, much of the MGB's charm lies in its mellow tractability and good low-end grunt, so I'm not sure that moving the power curve higher up the rpm band serves its better nature, except for competition.

The one performance modification that really worked was the installation of a rear anti-roll bar and a heavy-duty front bar. The ride is just slightly more jittery over normal road bumps, but the flatness and balance in corners are big improvements.

So early one frosty October morn, de-cammed, re-carbed and heavily sway-barred in the all-new $10,000 MGB, Chris Beebe and I loaded up for the 2000-mile round trip to the Runoffs at Road Atlanta. We'd made this trip in a variety of vehicles nearly every autumn for 20 years, and this was the last year the Runoffs were to be held at Road Atlanta. (This year they'll be at Mid-Ohio.) Time for one last sports-car run deep into the South, the land of retreating winter.

We loaded in the morning darkness and discovered, much to our satisfaction, that the MGB holds a lot of stuff. The trunk is surprisingly roomy, even with the spare tire sharing luggage space, and the well behind the seats can swallow up a tent, two sleeping bags, a camera bag and a car blanket in case of heater-fan switch failure.

Sunrise struck on the foggy Rock River as we motored into Illinois, then aimed southeast toward Indiana and Kentucky. Before its rebuild, the MGB had a standard 4-speed gearbox, making highway travel fairly noisy and hectic. I found a used overdrive box, however, which dropped the revs at 70 mph from almost 4000 to 3300, a huge drop in both piston speed and mental anguish. Another good mod. By nightfall we were actually 700 miles from home, in Chattanooga, just as you might be in a regular car.

Computations at the end of the day: 24 mpg, two quarts of oil, no rattles, red lights, untoward noises, bad smells, overheating or underheating. Amazing. And the car had a nice ride on the highway. Except for our out-of-kilter wheels, which made the steering vibrate like a low-pitched tuning fork at about 62 mph.

New seat foam in an MGB perches you fairly high under the low-slung convertible top, so you always feel as if you're wearing a tweed cap pulled down to your eyebrows. No sun visors needed. This is fine unless you're trying to see an overhead traffic light at an intersection. It's easiest just to wait until people honk or call the police; saves the neck muscles.

On Thursday morning, all those nights in the garage paid off. A warm, balmy autumn day in Georgia with full autumn color, winding roads through Dalton, Chatsworth, Ellijay and Dahlonega, top down, flicking the overdrive stalk in 3rd and 4th for just the right amount of growl and power in the curves and hills of the Chattahoochee National Forest. One of the best days of driving I've ever had, listening to the exhaust resonate, looking up into the tunnel of fiery red and sun-yellow trees.

The MGB handles remarkably well on these roads. After all these years of automotive progress, it can still pass itself off as a nearly modern sports car. With good ground clearance, firm but well-damped suspension that takes jolts without upset, a solid unibody chassis, roll-up windows, reasonably good heater and weather tightness, comfortable seats and nice gearbox, it may be the most livable and practical sports car the British ever produced.

One thing the MGB does not do, we discovered on this trip, is attract attention. Few people on the highway give it even a sidelong glance; patrons do not disembark from cafes to admire, as they did on our trip with an MG TC; gas station attendants do not say, "Nice car" or "What is that thing?"

Even though it is now a 30-year-old design, the MGB seems to have just made the styling cut into the modern era, where it is assumed to be part of the norm; handsome, maybe, but not flamboyant or unusual. While the MGB is idiosyncratic enough to qualify as an acquired taste, its appeal is broad enough to be understood and accepted without comment, like a preference for dark beer or the music of John Coltrane. Perfectly defensible.

And that, too, is part of the car's appeal. Its somewhat archaic mechanical underpinnings and taut, envelope sheet metal bridge the gap between old English charm and modern science, in much the same way a Jaguar E-Type does on a more complex level. You have to drive one to sense and appreciate its Queen-Victoria-Meets-The-Beatles personality.

So if you go on the road with an MGB looking for approval, it's best to look within. Abingdon made more than half a million of these cars, including the GTs, so people are accustomed to seeing them around. Only other MGB owners (or would-be owners) care very much whether you have one or not. Which is yet another reason many of us like the car; unsolicited admiration from the ill-informed and overstimulated is not a burden we have to bear. It's a quiet, low-key classic.

Top down on a fine fall evening, we arrived at our usual Lake Lanier campground near Road Atlanta. Camped, drank, smoked cigars, sat around campfires and went to races. The MGB looked good, catching flickers of firelight at the edge of the campsite.

On the way home we took the backroads to Chattanooga, coming in through Chickamauga National Battlefield, where two generals of only moderate genius, Rosecrans and Bragg, had their massed troops slug it out in heavy woods along Chickamauga Creek for two September days in 1863. Four thousand dead and 35,000 wounded. A gloomy place in the cold autumn rain now, as then, no doubt.

From there we drove past my old basic training grounds at Fort Campbell, Kentucky, also raining darkly as it was then. But in northern Illinois the clouds broke, with sunshine and a warm south wind—just as we passed a graveyard that said "Egan Cemetery" over the gate.

Finding no Egans buried there, I decided, as my forebears apparently had, not to stick around. We put the top down and drove into Wisconsin on country roads, car running fine, leaving a trail of exhaust resonance and swirling leaves.

We pulled into my driveway, having set a new personal (if not official) record in a British car: 2046 miles without making a single repair or adjustment. The toolbox had never been opened. Shop rags were still folded in the trunk; our hands were clean.

Three winters in the garage vindicated, just before another one. And only a week after Columbus Day, but well before Thanksgiving. ⊚

> *Finding no Egans buried there, I decided, as my forebears apparently had, not to stick around.*

Continued from page 139

push-handle on the back. You run behind it, then stand on the runners, gliding down the street, pushing occasionally with one foot. You can carry a child, another adult, a bag of groceries or a gallon of aquavit on the seat of the chair. People in Jukkasjärvi go everywhere on their *sparks*, which seem to cover ground almost as fast as bicycles.

These vehicles are practical in the north because the roads are graded, hard-packed snow, with a smooth, uniform base, like a well-groomed ski run. Also, with the sun shining low on the horizon for only a few hours a day, nothing ever melts, so there's no slush or ice. And no salt. The roads and countryside have the clean white look of fresh powder, theatrically perfect.

And ideal for the winter driving of Volvo's new 850 sedan.

The 850 GLT, as noted in our October 1992 road test, is something of a departure for Volvo, being its first front-wheel-drive car. Despite its traditionally boxy lines—but at least kind of handsome, *windswept* boxy lines—it's also lighter and sportier feeling than previous offerings from Göteborg, with a smooth, free-revving dohc alloy 5-cylinder engine of 2.4 liters. Torque and power are both remarkably good, output being 162 lb.-ft. of torque at 3300 rpm and 168 bhp at 6200 rpm.

The point of our driving the car in the Norrland, of course, was to test its mettle on snow and ice, which we did with a 330-km cross-country drive to the Sami village of Tarendo and back, and on a frozen lake one morning.

Volvo has gone to great lengths to make the 850 an ideal "snow car" for northern climes, equipping it with standard ABS and optional traction control and putting 60 percent of its weight over the driving front axles. The anti-squat, anti-drive suspension geometry—as well as the power-assisted steering—have also been designed for optimum road feel and feedback on slippery roads. The Delta-Link rear suspension is a rather elegantly packaged twist-beam trailing-arm system that helps steer the car, and the equal-length front axles are supposed to eliminate torque steer, and do.

You also have the usual winter Volvo perks such as heated seats (better than Vicks), heated mirrors and a cabin temperature-control system that keeps your critical parts—mostly hands and feet—warm without hothouse stuffiness. Volvo engineers note that when interior temperatures are raised from 70 to 80 degrees Fahrenheit, driver re-

sponse time is 30 percent longer and the number of totally missed signals in an alertness test increased by 90 percent. Even without glögg.

Some of our test cars were also equipped with a new gasoline-fired preheater for the cooling system, which circulates hot water through the interior heater and runs the fan on a timer system. Hot starts reduce engine emissions and wear, and prevent thousands of Swedish drivers from shouting, "Damn, I hate this climate!" on the way to work every morning.

On to the driving.

At moderate speeds on snow-packed roads, the 850 tracks through corners with excellent steering feel and just the

■ ■ ■ ■

Rather than being merely dark, the whole landscape takes on an enchanted, magical quality, as if lighted from within. Odd, but beautiful.

■ ■ ■ ■

right amount of body roll to give the driver an accurate sense of transitional bite. Pushed harder (or way too hard), it simply uses more road in a wider version of the same neutral arc. If a more sudden change of direction is called for, the car responds well to the pitch-and-catch technique: Throw the rear end out and let the front tires pull it back in line, shooting the car out of a corner.

All of this was aided, of course, by studded snow tires, which allowed us to experiment with snow handling at speeds that would normally have had us sliding sideways into Norway. Sweden, Norway and Finland still allow studded tires, but new legislation limits the weight and number of studs, to cut down road damage. So Volvo has come up with a new type of tire, work-

ing with Goodyear, Michelin and Gislaved, which uses metal-cored plastic studs. Easier on the road surface than the all-metal type, and quieter. In any case, they really work.

I spent most of my time in a 5-speed manual car, but Volvo also has an automatic transmission with Economy, Sport and Winter settings. If you select Winter, the car starts off in 3rd gear, so you don't do an accidental doughnut out of the driveway and hit your neighbor's *spark*. That, with the optional traction-control system (TRACS) that brakes a spinning front tire at speeds up to 25 mph, makes getaways undramatic.

All this engineering and the studded tires conspire to make a car that can be driven very hard on snowy roads and requires either an act of God or inspired genius to put it off the road. But it can be done. We had a large group of Portuguese journalists on the trip, and five or six of them managed to bury their 850s in snowdrifts some distance from the highway. But then snow driving is not a folk art of Portugal. I am proud to say that the American crew never put a wheel wrong, so there is finally some payoff for having lived through many winters.

The day we left Kiruna, we waited for our flight at the small terminal. From our waiting room we could see down one of the corridors of the airport office building. Suddenly the lights in the hallway went out.

Power failure?

No.

A door opened up and angels did appear. Actually, they were Swedish schoolgirls wearing white gowns and crowns of small electric candles in their hair. Each one carried a lighted candle in her hands. It was the pre-Christmas festival of St. Lucia, and the children were visiting all the hospitals, old folks' homes and public buildings in town.

They marched slowly through the airport and down the darkened hallway, past the offices, singing traditional Christmas carols in complex harmony that would do credit to a professional children's choir. Clearly, they had a very good music teacher. Along with beauty and innocence, and a refreshing isolation from the world of our own local schools, which are just now grappling with crack and handguns.

It was a moving sight and a memorable last-minute Christmas gift from Sweden, which I hope never entirely loses its middle way. ◉